THE GENESIS EFFECT™

∞∞∞∞∞∞∞∞∞∞∞∞

Spearheading Regeneration with

WILD BLUE GREEN ALGAE

∞∞∞∞∞∞∞∞∞∞∞∞

Volume 1

∞∞∞∞∞∞∞∞∞∞∞∞

by

Dr. John W. Apsley, II

∞∞∞∞∞∞∞∞∞∞∞∞

Published by Genesis Communications, LLC

THE GENESIS EFFECT

Spearheading Regeneration with Wild Blue Green Algae

by Dr. John W. Apsley, II

Published by:

Genesis Communications, LLC
4000 Mitt Lary Road
Northport, AL 35475

First Printing 1996
Printed in the United States of America

Library of Congress Cataloging in Publication Data
Apsley, John W., II
The Genesis Effect: Spearheading Regeneration with Wild Blue Green Algae

Bibliography: p.
Includes index.
1. Blue Green Algae—Handbooks, manual, etc. I. Title.
2. Ribonucleic Acids—Handbooks, manuals, etc.
3. Regeneration—Handbook, manuals, etc.
4. Clinical Nutrition—Handbooks, manuals, etc.
5. Nutritional Therapeutics—Handbooks, manuals, etc.

CIP (Applied For)
ISBN 0-945704-01-1

Publisher: Linda K. Carroll
Cover Design: David Doyle

FORWARD

Everyone wants to attain and maintain good health. The information contained in this book written by Dr. Apsley, will help many individuals accomplish these goals. However, most people do not stop to realize that doctors are really only repairmen. Individuals are the ones rightfully responsible for the health and maintenance of their bodies. This book encourages individuals to look at themselves in relation to their biosphere and how they are interrelated.

We are all grateful to Dr. Apsley for writing this revolutionary book. His unique explanation of how to utilize blue green algae will prevent a great deal of sickness and suffering and will improve the lives of many plagued by chronic fatigue, insomnia, depression, anxiety and numerous other physical and mental conditions.

Blue green algae is not "new," but it is becoming increasingly better understood. Although much of our medical and scientific community is still lagging ten to twenty years behind in the needed research, progress is being made. Many patients who are mistakenly diagnosed as being neurotics or psychotics, or as having ADD, hypoglycemia and numerous other syndromes are actually suffering from nothing other than nutritional deficiencies. Their health problems are often miraculously solved by simple dietary changes including *The Genesis Effectors*.

This book will become indispensable not only for the doctor helping to translate the complexities of *The Genesis Effect* and blue green algae into the means for achieving better patient care, but also for individuals who want to have better health. On the basis of my own experience, I can attest to the accuracy of Dr. Apsley's approach of utilizing blue green algae for a healthier American population.

John L. Stump, D.C., Ph.D.

ACKNOWLEDGEMENTS

I wish to express my sincerest thanks and gratefulness to the many named and unnamed individuals who made these series of books possible. It has been a long road.

First, let me give my thanks to Dr. William Powell Cottrille, my personal physician and teacher, totally responsible for starting me on my journey as a health care provider. Next, I wish to thank my chiropractic mentor and teacher, Dr. Daniel Duffy, Sr., whose patience and persistence were the only reasons I kept my mind open long enough to finally have basic realizations about what the subluxation (chiropractic) paradigm was all about. To my dearest friend and tutor of life and death, S.N. Goenka, whose devotion to the wholesome activities of life, to mental discipline, awareness and equanimity gave rebirth to my faltering ways. To my parents, who struggled all of their lives to spare me from many of the encumbrances of daily life so that I might someday accomplish my fondest dreams and aspirations. To the best team of publishers of all times, the staff at Genesis Communications, for their belief, trust and hard, world class work and moral support edging me on to reach my final goals. For author Anna Bond's unrelenting encouragement and persistence to achieve my fullest potential with Super Foods. To Dr. Dale Burgess, for giving me a place to systematically work out the best regenerative approaches and protocols under some of the most difficult clinical settings. To my daughter for remaining one of my most staunch supporters and believers in what my life is all about. And lastly, to Miriam, my mate and partner, whose devotion has given me the courage to face my most difficult challenges and responsibilities in life.

My most heartfelt thanks and gratefulness must go to my publisher, Linda Carroll, untiring editor and trusted colleague, for without her expertise and guidance, *The Genesis Effect* series would not have been possible. Linda has not only added breadth and dimension to these works, but has indeed been the creative genius behind their approach, delivery and structure. To all the readers of these volumes, know that she is at the heart of our presentations.

May all beings be happy!

PREFACE

Contained in this book will be nutritional knowledge that you must push yourself to explore. At first, it may not come very easily. If you are a reader only beginning to look at nutrition, this is an excellent place to start. If, on the other hand, you are an old pro at picking up various books on nutrition, this book is bound to change your mind about many things you thought were good nutritional practices.

To begin with, I have always adhered to the principle that **the proof is in the pudding!** A nutritional book that does not deliver is useless. Therefore, starting with this book, and then with additional volumes to soon follow, you will be challenged to explore this information on at least three levels. If you really apply yourself and seek the counsel of a properly-qualified "wholistic" health care practitioner, your rewards should be dramatic because you are bound to be successful!

As stated above, you will be challenged to read this book at three main levels of awareness. The first level of understanding is based on faith. Faith is a necessary ingredient building toward the other two levels of knowledge. When you are first applying any information contained in this work, you are just starting out. So, indeed, you must have a degree of faith to start incorporating this information. I ask you to not go overboard and become blindly faithful about any of it. Rather, give it a fair and honest trial with your eyes wide open, evaluating it by the other two levels of understanding (intellect and experience).

Next, give intellectual understanding to what you are reading and incorporating into your life from this book. This is the second level. A good intellectual approach is to note what type of footnotes this book contains. Are they reliable footnotes that back up the points being made? Also, does good 'ole common sense seem to be prevalent throughout the text, or is there foolishness? Is this work well organized and practical? Use any type of intellectual evaluation of this kind for this book that you wish, just keep an open mind. And whatever you do, do not become dogmatic about it either, even if you do agree with most of it. Intellectual dogmatism about nutrition and health is what has gotten us into so much trouble already. It has allowed most of the important works on *The*

Genesis Effect (see Chapter One) to not only be largely ignored, but even in some instances, to be condemned without any justification (other than it simply isn't popular right now).

But at this point, your job is far from over. You must acquire the third level of understanding about this book—which is the most important level of understanding. You must **experience** for yourself just what this book is all about. The first two forms are **pre-requisites** to the experiential level. Without the experience of these health practices, **no insight** can come about. Without insight, there is **no wisdom**, and without wisdom new ways of understanding cannot be found. Then, it soon becomes a descent into dogma and blindness. This is how any book on nutrition ought to be read. To gain experience with the information contained in this book, you will find it useful to add liberal amounts of wild-grown blue green algae to your own diet.

Remember, use faith in reading this book, but not blind faith. Use your intelligence in sifting through its pages, but leave dogma behind. And most importantly, use the first two to gather your own experiential knowledge on *The Genesis Effect*. Develop clear insight and wisdom over the contents of these pages so that no one can take away from what you have truly learned and experienced for yourself.

Be sure and watch for the following commonly-experienced traits of blue green algae consumers: (*a*) increased hair growth and thickness; (*b*) faster and stronger growing nails; (*c*) fading of wrinkles and (liver) age spots; (*d*) increased energy and improved weight profile; and (*e*) better mental happiness and emotional well being.

The first three chapters of this book are for everyone except the skeptic scientist. Chapter Four is for the skeptic scientist, and Chapter 5 is for everybody, including the skeptic scientist.

The reader is advised to read this book in the following way:

1. Briefly scan the Table of Contents and then thoroughly read the Preface.
2. Read in sequence the quote at the beginning of each chapter followed immediately by the last page of each chapter under Genesis Quicknotes. In this way, you will be best prepared to fully become empowered with an exciting new paradigm of how to bring regeneration into everyone's life.
3. Now read the book cover to cover.
4. Set the book aside for one full week, and then reread each chapter's initial quote followed by the Genesis Quicknotes. Continue with this ten-minute weekly review for the next four weeks.

5. Become emissaries of regeneration for yourself, your family, your friends and the world. Let the healing begin today!

Special Reader Advisory

This information is provided under the constitutional law of the land. Additionally, the reader is strongly encouraged to "balance" all pro-health statements contained herein with the fact that, to the best of the author's knowledge, the Food and Drug Administration (FDA) has not evaluated any of the statements made in this volume. And finally, the information within this book relating to health and wellness must not be construed by the reader to be making any type of claim as to the cure, mitigation, prevention or treatment of any medical disease. Thus, health improvements resulting from the information contained in *The Genesis Effect* series and the use of super foods are to be regarded as nutritional only.

Nothing can be more important than your health, so use this book wisely. Your education of the regeneration paradigm (i.e., *The Genesis Effect*) should be impartial and unbiased. Feel free to discuss the information in this book with your doctor. In order to control your own health destiny, you, the reader, must take full responsibility for your health and how you use this book. In making the leap from being a victim of circumstances to becoming self-actualized in your health status, you must follow in the footsteps of all those before you. The path is universal, the challenges are universal. Chronic degenerative disease is the same in either sex, the same no matter what your racial heritage, the same whether you live in Russia, China or the United States of America. You will face the same challenges as those who have gone before you, and you will successfully overcome each hurdle as they did – one step at a time. Do not accept failure in your journey, but feel free to take rest from your efforts whenever and wherever needed. And remember, too, that in your times of trial and tribulation, the tools will appear if you have faith, understanding, insight and persistence.

Please note that pregnant women are especially urged to consult with their physicians before using techniques described in this book. Also, do not substitute this book knowledge for the advice and treatment of a physician or other licensed health professional. This information is fully compatible and can be used in conjunction with professional care. With the aforementioned all stated and understood, the author and publisher expressly disclaim responsibility for any adverse effects resulting from your use of the information contained herein.

Special Advisory for the Scientific Minded

My research on the subject of regeneration started over 17 years ago. After searching long and wide, first at Michigan State University, and then later throughout the east coast stretching from Boston to Miami, I came to realize that most of the applicable regeneration information useful for clinical practice was written in the first half of this century. Indeed, even today it is difficult to find literature or documentation that contains information useful for an updated bibliography. However, the search does go on, and I invite any reader who might possess such documentation to please send it my way for inclusion in future editions of this and other *Genesis Effect* volumes.

Additionally, the older references should be placed into their respective contexts to judge their merits and shortcomings. While it is true that 2+2 equaled 4 in 1950, and that 2+2 still equals 4 today, most academicians all too readily dismiss well-done studies performed more than ten years ago. This is akin to saying that no one should place much credence into today's (1995) well done studies because in 2005, they will be regarded as useless anyway!

Furthermore, David Eddy, Professor of Health Care at Duke University, recently undertook a review of the modern basis to medicine and determined that all medical techniques in common use today were indeed only 15% confirmed by scientific studies published in peer-reviewed journals stretching out over the past 20 years! The General Accounting Office of the U.S. Congress confirmed this finding. However, the Berkeley study went on to say that of these studies, only *one and one half percent* (1.5%) of all articles published in medical journals were scientifically sound. "Many treatments have never been assessed at all." Eddy, a highly respected consultant to the Agency for Health Care Policy and Research (AHCPR) stated that for the majority of areas in medicine, the scientific evidence in support of today's current medical practices is "between poor and none. . . . Usually the *best* evidence was something less than a randomized controlled trial." Curiously, in many areas where there were good scientific studies, they often contradicted current practices. The *British Medical Journal* concluded that, "The weakness of the scientific evidence underlying medical practice is one of the causes of the wide variations that are well recognized in medical practice."[1]

And lastly, it must be mentioned that although this author openly invites any reader to submit good current science that confirms or denies the premises of these *Genesis Effect* volumes; for the record, the older scientific community entertained the exploration of the regeneration paradigm with an openness only rarely expressed in today's central constructs heavily dominated by reductionism. Have things today gotten so reductionistic that in our attempt to explore scientific thought, we have abandoned time honored non-reductionistic methods? For example, one great scientist of this earlier era, Linus Pauling, told the medical community categorically in the 1970's that for humans under stressful living conditions, 10,000 milligrams of vitamin C daily was necessary in order to maintain health. The medical giants laughed openly. Yet many of these medical doctors – all having an average life expectancy of 58 – were long dead by the time Pauling succumbed to his ten-year battle with cancer at the ripe old age of 93! Only the last few months immediately prior to Pauling's death were not top quality productive years. Obviously, he knew something beyond the understanding of modern reductionist thinkers. Only recently have UCLA (1992) and other recognized scientific strongholds begun to confirm his work. It may be too late for most of us if we only allow reductionistic thinking to determine our present health paradigm. So, why not include other forms of good science, open-mindedness and best case scenarios to augment our accepted methods of analytical thinking today, especially on the subject of regeneration?

For a lively and most interesting discussion regarding the legal ramifications of educational materials on nutrition and health, the reader is strongly encouraged to see Appendix B.

May your journey be most successful... Godspeed to you all!

Table of Contents

Chapter One

The Genesis Orientation

The human body has one ability not possessed by any machine—the ability to repair itself.

George E. Kriley, Jr., M.D.

Introduction

Wake up now, pay attention – this book will dramatically change your life. Read diligently to profoundly help yourself, your family and your friends find health secrets that should never have been secrets at all. And above all else, put time and energy into this most precious study, so that you can enjoy the full health benefits of regeneration for yourself.

Make no mistake that learning and experiencing regeneration firsthand are hard work. *The Genesis Effect* is the how, why, where and what of regeneration. You have today a unique and conveniently available wild, whole super food that is full of the most potent regenerative nutrients known to humankind. Its name is blue green algae. In reading the story of *The Genesis Effect* and blue green algae, you are embarking upon the most important discovery of your life: how to live long, how to stay healthy, how to prosper and how to alter the course of our planet's history.

The Genesis Effect

The Genesis Effect is what lies behind the normal healthy growth of an individual as well as *all* healing processes going on within the body. Therefore, *The Genesis Effect* is also the causal force behind all regeneration. *Webster's Dictionary* defines regeneration as "to endow with new life and vigor."[2] However, it also means, "The renewal or reproduction of cells, tissues, etc., in the ordinary vital process."[3]

Nutritional *Genesis Effectors*

Genesis Effectors, as discussed in this book, will center around uniquely powerful nutritional substances found only in therapeutically useful amounts in super foods. We will further define *Genesis Effectors* to include nutritional complexes that:

1. Serve as the basis for all regenerative actions.
2. Induce unscheduled regeneration when needed.
3. Supply the special nutritional needs for **optimal function and growth**.

Nutritional complexes that inspire regeneration and optimal health are almost completely overlooked or described only vaguely today in the health-conscious literature. Now, for the first time, this book offers a precise and close look into the world of super foods containing regenerative and optimal health-inducing nutrients.

The Genesis Effect, to be optimal, also requires:

1. Access to plenty of pure, clean, unpolluted water and air to drink and breathe.
2. A calm and peaceful environment.

Today, toxic chemicals abounding in our everyday environment clog our vital metabolic pathways. These toxins accumulate slowly over time, thus avoiding detection. These poisons must be detoxified from our tissues for *The Genesis Effect* to adequately blossom within.

Understanding the adversaries of *The Genesis Effect* discussed in this book is equally important. These are:

1. Heavily processed, refined and preserved foods.
2. Highly-cooked fat foods, especially of animal origin.
3. Today's modern agribusiness.

These three factors characterize most of our Standard American Diet (SAD) and most of the standard chows our animals are fed.

How Do We Regenerate?

Humans and animals regenerate at the cellular level. **This means that proper optimum nutrients (i.e., Genesis Effectors)[a] must be present**

---

[a]*Genesis Effectors* are cybernetically expressive. Cybernetic systems within our bodies are physiologically open-ended systems, as opposed to the closed systems biologists are

inside, as well as outside, of the immediate cells and tissues of both humans and animals.

Hence, the ability of cells and tissues to follow through with a perfect replication, creating vital new cells, is for the most part dependent on: (*a*) what type of raw materials the original cells have to work with; and (*b*) some of the original cells remaining intact and unscathed. Collectively, this process dictates human longevity as well.[b]

Since both humans and animals possess trillions of cells, each with their own minuscule reserves, internal reserves can be called upon for use in times of deficiency. However, this habit of going to the bank to borrow to pay the current bills without making any deposits never seems to stop. Sooner than later, this is how degeneration creeps in.

How Come We Have Never Heard of *The Genesis Effect* Before?

This is a good question. Actually, the knowledge regarding regeneration has been around for some time.

What it basically boils down to is that *Genesis Effectors* are highly perishable complexes found only in raw foods. Therefore, they threaten profits because of the spoilage risk. **These factors are usually destroyed by the various methods used for preserving our food, and your health pays the price for it!** Additionally, economics also dictates that farmers grow produce as cheaply and conveniently as possible, and this, too, affects the quality and quantity of *Genesis Effectors* found in our groceries.

By the same token, when grants for research dictate in what narrow and limited areas to apply good science, we often get a distorted view, at best, of what mother nature is all about. It is not economically feasible to patent carrot juice, high in a cancer preventative known as beta carotene. Rather, we try to find ways to patent a money-making process for syn-

---

accustomed to studying. Closed systems are limited systems. They can perform only simple tasks as compared to regenerative tasks. Closed systems tend toward linear processes; whereas, open systems are non-linear, integrating a plethora of dynamics on a constant basis. Therefore, any therapeutic approach that uses a linear model as its thrust, a this for that, a "here take this drug for that disease," becomes totally and definitively obsolete in health conditions caused by breakdowns in the open-ended, multifactorial processes of life. In essence, *Genesis Effectors* are the supreme commanders of our cybernetic systems, and are thus the determinants of all regenerative phenomena.

[b]For a detailed synopsis on extending the human lifespan with blue green algae, see *The Genesis Effect: Slowing, Stopping and Reversing the Aging Process* in Appendix A.

thesizing beta carotene from cheaper sources. "For what shall it profit a man if he gaineth the whole world but loseth his soul?" (Matthew 16:26).

Another example of the resistance to understanding the regenerative qualities of foods is found in the farming industry which is taught that more is better. Farmers are not taught how to maximize the *quality* of their produce so much as the quantity. Then, to back up these same technologies, grants are again incorporated to offer token studies that fool the public into thinking that our food supply is just as nutritious as ever. What these studies would like to overlook, but this book will not, is that if everything were so much better today in how we grow our foods, then why is the average wheat berry only 6% to 9% protein today as compared to 14% to 17% protein forty years ago?[4,5,6] There are even reports that at the turn of the century, America was growing wheat containing up to 40% protein. This same trend has extended to other grains as well, such as corn.[7] Additionally, in the late 1940's, a typical bowl of spinach contained 154 mg of iron. Today, that same amount of spinach will contain less than 1 mg of iron! (See Figure 5.1, page 40.)

These are just a few examples of how we are somewhat misled by our current scientific way of doing things. Another example of our need for *The Genesis Effect* was illustrated by Albrecht. He found fatter animals and less muscular animals eating off lands whose soils were depleted by heavy farming and pesticides because the farmers were growing less and less proteinaceous (protein-containing) grains. These same animals had more diseases, and in many cases, failed to reproduce normally.

Many men who were raised on meat from these animals flunked the Army's physical examination for induction into the service at higher rates than normal. They also showed dramatic increases in dental diseases and other bone deformities.[8] I can't help wondering if this is also related to why there has been such an increase in obesity in people as well. Today, our population is replete with people who crave salt, sugar, alcohol, tobacco, greasy foods, caffeine and drugs of all types. These cravings have a cause and a cure. Albrecht called this the *hidden hunger* phenomenon. I'll give you a great illustration of this.

Albrecht's *Hidden Hunger* Phenomenon

In the fall of 1978, I travelled to Forres, Scotland. While there, I planned on visiting a very special place called Findhorn where some remarkable things were happening. When I arrived, I met people who had lived there for some time. An older couple caught my attention, and I dined with them on several occasions. They looked like they were in their late sixties;

however, they were quite active and full of spirit. I eventually started asking them about their life, and they told me they had lived in the Findhorn community for only five years. Before that, they had lived in the adjoining trailer park where they had retired.

They showed me a picture of the two of them from the days before they joined the community. They looked like they were much older than the robust people I saw. The wrinkles were clearly evident in the photograph. Yet, their skin was now soft and smooth. And the rosy tint in their cheeks was too much. They looked like the perfect "postcard couple" you want to buy to send home!

Then, they told me exactly how old they were. They were both over 95! I didn't believe it. Finally, they had their friends come over and swear to it, plus they confirmed it with their British drivers' licenses. I was astounded. What had turned these two around?

A few days later, I was working hard in one of the many organic/bio-dynamic gardens that abounded in the community. One of the tomatoes was ripe, so I picked and ate it. To my astonishment, my whole mouth lit up with its overwhelming sweetness! Never in all of my life had I tasted such a delicious tomato! I was picking these tomatoes for transport by way of a truck nearby. In this truck were the weekly supplies for the meals, and some of the produce came from local farmers who also grew their produce naturally. I slipped into the truck and relieved it of one of the tomatoes so I could eat it. I had no plan in mind, I was simply still a little hungry. When I bit into it, it tasted like pure water! The taste from the first tomato was still fresh in my mouth; not more than two minutes had passed. And so, when I tasted this second tomato, the difference was once more astounding.

I learned that the more bio-available the minerals are for the plants, *the sweeter the vegetables are*, because this is where *Genesis Effectors* are to be found. We should always be eating these sweet vegetables. The second tomato is a great example of what causes the *hidden hunger* described by Albrecht, **even though it had been organically raised**. I knew instinctively that eating anything less than the first tomato was inadequate to maintain total health. The satisfaction in both its mental rewards as well as its physical rewards was indescribable. In a sense, I had always been "craving" such produce, only I didn't know it until I stumbled across it for the first time in my life in Findhorn! This was my first experience with a true super food.

In this example, the older couple had been eating inferior produce for years, leading them to suffer from *hidden hunger*. They aged just like most people do today, never realizing that this need not be the case. Then, when they got into the Findhorn community, they ate almost exclusively from the specially-grown super produce of community gardens. It has become very clear to me, based on my own experience with this *hidden hunger* phenomenon, that properly-processed blue green algae satisfies the *hidden hunger* plaguing many people today.

Conclusions

The more the food industry is motivated by profits, to that same degree, our foods will proportionately produce the *hidden hunger* phenomenon. There is no amount of money in the world worth sacrificing *The Genesis Effect*. There can be no justification or excuse for the motivation operating our food industry. For example, when flour is bleached, over 40 trace minerals and vitamins are removed. Yet, only seven synthetic nutrients which are added back in, are expected to make up for this loss. Another example is synthetic vitamins that are inherently incapable of eliminating, in part or in whole, the *hidden hunger* phenomenon.

One more thing can be said about *The Genesis Effect*. **Whenever man tries to copycat mother nature for purposes of greater profit, the cost will be in the loss of *The Genesis Effect* to the same degree.** The only possibility that man has regarding *The Genesis Effect* is to offer assistance to mother nature by first obeying her rules. Two of the greatest 20th century nutritionists, Dr. Weston Price and Dr. Granville Knight, stated: "The laws of God and nature are immutable. They cannot long be broken without retribution. . . . **Life in all its fullness is mother nature obeyed**."[9]

Genesis Quicknotes

1. *The Genesis Effect* is caused by the body receiving optimal nutrients from ideally-preserved, raw super food sources that initiate repair and regeneration within the body.
2. Optimal healing via *The Genesis Effect*, must first be prefaced with detoxification, and next with optimal delivery of oxygen to all tissues.
3. A calm, low-stress environment is essential in order to provide the framework under which *The Genesis Effect* may firmly establish itself.
4. The enemies of *The Genesis Effect* are contained abundantly within processed foods and modern agribusiness practices.
5. Regeneration or degeneration is a "cumulative" process, the consequences of which our children will bare.

6. Since commercial production often replaces the wholesomeness in our foods with profiteering, *The Genesis Effect* has never been economically or politically popular in this country.
7. Soil microorganisms are the most important factor in both crop nutritive values and preventing the *hidden hunger* phenomenon.
8. The *hidden hunger* phenomenon is manifest in direct proportion to the soil's deficiency from its optimal state. The prevalence of this *hidden hunger* has brought nothing but suffering and misery to humankind, and therefore must be eliminated.
9. The more nutritious the vegetable produce, the sweeter it will taste. Therefore, one test for evaluating the content of *Genesis Effectors* in produce (especially in vegetables) is to taste "how sweet it is."

CHAPTER TWO

SOIL FERTILITY AND *THE GENESIS EFFECT*

Diseases are created chiefly by destroying the harmony reigning among mineral substances present in infinitesimal amounts in air, water, food, but most importantly in soil. If soil is deficient in trace elements, food and water will be equally deficient.

Peter Tompkins and Christopher Bird

If you take a farmer who uses synthetic chemical fertilizers and compare his/her produce to a farmer's produce who uses all natural fertilizers (organic/bio-dynamic techniques), the first farmer is said to be able to grow 60% more produce per acre than the second. The common synthetic fertilizers on the market today are typically comprised of nitrogen, phosphorus, potassium and calcium (NPKCa for short).[10] Although greater crop yields (quantity) have been thought to result from the use of these synthetics on soils, the nutritional value must sooner or later diminish. According to the same study, control plots that received no fertilizer of any type produced 13.7% more calcium and 28.8% more iron in the vegetables.[11]

All farmers are taught to think that greater yields per acre make more sense economically. Ebeling stated the following about these findings:

> However, in some extensive and well-conducted field experiments in which crop productivity could be increased, the normal nutrient content in food crops generally could not be appreciably improved by the addition of fertilizers under practical horticultural practice. It appeared from these experiments that the

> nutrient composition of a plant is more influenced by inherent soil properties, such as chemical composition, microorganisms and physical structure, as well as by climatic environment, than by any change brought about by fertilization.... In any case, the agronomist and the farmer are rarely concerned with the nutrient composition of a crop. Varieties, cultural practices, and fertilizers are all decided on the basis of yield (quantity) expectations.[12] (parentheses added)

Michigan State University conducted a ten-year experiment to see how a synthetically-complete fertilizer containing only nitrogen, phosphorus and potassium would affect the nutritional value of corn, soybeans, wheat, oats, oat hay, brome grass hay and timothy as compared to very depleted (exhausted) control land. The results were that all of the plants grown with the complete fertilizer showed no nutritional improvement in any category except for minor improvements found in the timothy crop![13] Thus, corn may grow for many years in depleted soil (using any amount of synthetic fertilizer), and still be totally unfit for human or animal consumption.

There are many examples in the literature suggesting that if soils are depleted in specific trace minerals, then some plants will show an improvement in their nutritional value when these trace elements are added to the standard complete NPKCa fertilizer.[14] Other experiments have shown conflicting evidence, which has led the researchers to conclude that the addition of trace elements to the soils does not appreciably influence nutritional values![15,16] Thus, plants cannot evenly assimilate minerals disporportionately provided by scientific methods. However, rock dust's inherent mineral ratios provide the ideal proportions for plants.

Natural Soil-Building Techniques versus Chemical Treatment

The truths spoken over 40 years ago are just as true today as they were then. Take, for example, an excerpt from an article written by Sir Albert Howard on the subject of natural versus man-made chemicals:

> It is always good to see the differences between natural and laboratory products emphasized, in recognition of the imponderable elements with which Nature endows substances, which can by no scientific skill be added to the synthetic product. The case in point is that of nitrates, and the report emanates from one of the United States' universities. It states: "natural nitrates have something that the artificial lacks, and there is no completely adequate substitute for it in the field of agricultural fertilizers. Chilean nitrate contains small amounts of vital impurities such as magnesium, iodine, boron, calcium, potassium, lithium and strontium, which are to plants what the vitamins are to human beings. It has been found that natural nitrate does something that makes apples stay on trees; that it does something to corn that results in better livestock fattened on it; that

chickens raised on nitrated-feed lay better eggs of greater fertility. **It is just as impossible to make artificial nitrates that duplicate natural nitrates as it is to make synthetic sea water that contains all the elements of natural sea water."**[17]

Some Answers

The natural farmer, who is a non-chemical user, whom we are led to believe may only grow 40% to 50% of the quantity of a synthetic farmer, knows that his produce may be twice as nutritious or more! Nutritional factors such as protein yields, vitamin quantity and enzyme yields would be even more startling if organic/bio-dynamic techniques for fertilizing were incorporated along with liberal amounts of rock dust. Rock dust contains an abundance of fully assimilable essential and non-essential trace minerals for all life on this planet that are highly bio-available to soil microorganisms. There are some excellent studies that clearly demonstrate this.

Let me cite one recent example stated by authors Jesse Stoff, M.D. and Charles Pellegrino, Ph.D.:

> During a lecture at the Massachusetts Extension Service, we heard about an important and timely study by the U.S. Department of Agriculture. While it is true that plants grown on synthetic fertilizers tend to be bigger and heavier, this is merely because they retain more water than plants grown without synthetics. **When all the water is evaporated off and their dried remains are analyzed, grains, fruits and vegetables grown on synthetic fertilizers contain significantly lower concentrations of protein, vitamins and trace elements (particularly zinc and manganese) than those harvested from plants grown by organic gardeners or simply allowed to grow wild**. We do not use the word "significant" lightly; in many cases, the difference is more than 50 percent. This means that the vegetables in your local supermarket are not only twice as big and twice as expensive as those grown before the "green revolution" (the widespread use of synthetic fertilizers), but you'll probably have to eat about twice as many of them to get the same nutritional value.
>
> Interestingly, many of the nutrients found in decreased amounts in synthetically-fertilized vegetables are now being revealed as being essential to the intracellular metabolism in fighting off a variety of disorders and diseases. Thus, eating "good" foods alone will not ensure adequate levels of vitamins and minerals anymore.[18] (parentheses and emphases added)

How come we don't read about this in the daily newspaper? Murray wrote an excellent piece on part of the reason why we do not hear very much along these lines:

> During a food-science convention in Dallas, Texas, about 1958, the late William Albrecht, Ph.D., Emeritus Professor of Agronomy, Department of Soils,

> University of Missouri, reported **then**, "Over 40 years ago, worldwide samples of soils (yielded a) viable protein output of 12%. **The minimum necessary for animal and human health being . . . (25%)**. In the United States, the samples averaged not 25%, not 12%, but 6%."
>
> During the late 1960's or early-to-mid 1970's, the U.S. Department of Agriculture released figures which indicated the average U.S. soils had a viable protein factor of (only) 1 1/2% to 3%. However, when educator and naturalist V. Earl Irons stated that the U.S. soils were incapable of supporting health because they were devitalized, **he spent a year in prison and paid several thousands of dollars as a fine.**[19] (parentheses and emphases added)

Only optimally-grown crops[a] will contain an abundance of *Genesis Effectors*. Fertilizing with rock dust and incorporating organic/bio-dynamic techniques can rapidly rebuild run down soils, both in the soils' viable protein output as well as in all other factors affecting the crops' nutritional value.

Soil Quality as an Indicator of Nutrition

As far back as the 1930's and 1940's, one of America's greatest agricultural scientists, who has already been mentioned, Dr. William Albrecht, demonstrated the differences between optimal versus standard soils. Unfortunately, Albrecht did not take his research a step further by making comparisons with the best-fertilized soils of his day to organic\bio-dynamic-raised soils complete with a plentiful supply of rock dust. These techniques were not really known in his day like they are today. I quote from a forward written in 1970 by Granville F. Knight, M.D., for perhaps the most important book ever printed on regeneration and degeneration:

> More recently, the increasing availability of artificial fertilizers of high nitrogen content has enabled the grower to harvest one crop after another without allowing the land to lie fallow—a custom which encouraged the multiplication of soil organisms that, in turn, would release soil nutrients as needed by plants. Often against his better judgement, the modern farmer has been forced to use monoculture, artificial fertilization, pesticides, herbicides and mechanization in order to keep ahead of ruinous taxation, inflation and ever-increasing costs of production. The result has been production for "quantity" rather than "quality," and the gradual destruction of our precious top soil and mineral reserves, in or beneath the soil. This has been well documented by Dr. Wm. Albrecht of the University of Missouri. Our markets are flooded with attractive, but relatively tasteless, vegetables and fruits. The protein content of wheat and other grains has steadily declined; this being a reliable index of soil fertility. Animal foods such

---

[a]Optimally-grown crops refer to crops grown using organic processes and in soil rich with vital nutrients from rock dust abundance.

> as fowl and meat reflect similar changes. Fowl are usually raised in cramped quarters and their food limited to that prescribed by man. As a result, cirrhotic livers are common and egg quality is inferior. Both groups are frequently treated with antibiotics, anti-thyroid drugs and hormones which produce castration, myxedema and waterlogged tissues. These practices are designed to stimulate more weight gain on less feed. The advantages to the producer are obvious; to the consumer, they are indeed questionable.[20]

Albrecht created many experiments to demonstrate that soil quality was the key to preventing degeneration. He took two different soils, one that was fertilized with synthetics, the other by his more complete methods, and placed them in trays, side by side. He next placed a vine-growing seed in each and let them grow. As the plants grew, their respective vines intertwined with each other. Eventually, an infestation occurred on the plant growing on the chemically-treated soil, **but not on the naturally-treated soil**. It was most interesting to follow the infestation on the infected plant, because the healthy plant's vine encircled it with many a twist, yet its resistance was too high to get the infestation. If the immune systems of the plants were so different, which soil would you rather have as the source for your food?

Albrecht also did similar experiments with soybeans and spinach, with the same results. He attributed his findings to the higher trace mineral content of the soils that were not replaced by the chemical fertilizers of his day. But, he also said that the soil itself released these nutrients as the organisms present in the soils freed up these trace minerals for the plants.

To help you better understand the idea that chemically-treated soils have very few soil organisms present as compared to naturally-treated (i.e., organic/bio-dynamic) soils, I would like to share the following story. I was once a member of an Organic Farming Club in Michigan. One farmer related a beautiful and germane experience to me. He said that the past season had brought with it a certain deadly bug infestation that had Michigan State University officials very worried. The infestation was spreading to many farms in the area, heavily destroying the corn crop.

This bug was usually a pest on cotton crops in the deep south, but had somehow migrated and changed so as to become a major threat to Michigan's corn crop. This farmer grew his organic/bio-dynamically raised corn around several other farms using chemical treatments. He, the organic farmer, was not experiencing any loss of crops at all, even though his neighbors who used chemicals were. Then, the state came in and insisted that everybody spray their crops with an insecticide. The organic farmer said no, but the officials were insistent. Then, they all took a walk on his land to check things out. What they found was even more surprising.

At first, they found many of the bugs all over his property, and then concluded that he, therefore, must spray. But then this farmer said, "Take a real close look." When they really studied what was going on, these scientists found that although the bugs were on his corn, **none of them were eating his crops**! There they all were, just sitting on his corn without eating. They flew off his corn and onto one of the surrounding chemically-treated farmlands to eat! I probably wouldn't have believed him if I weren't actually standing there with my own ears listening while he and a university official both confirmed the story!

The notion that a farmer using organic/bio-dynamic techniques would grow less is now seriously in question. Recently, studies comparing chemical versus natural means to raise crops were conducted by the National Academy of Sciences, and confirmed by Henderson at the University of California. They surveyed 14 different farms from across the country. What they found was that although it was a little more work to go the natural way, percentage yields were **as good as the yields from the chemically-treated farms!** They also found that it was ***cheaper*** to go the natural route, but that chemical farmers have lost their know-how to do it all naturally![21] In my mind, there is no excuse for not cleaning up the environment as it relates to agribusiness. But, oh, what a fight the chemical interests will wage, especially via their advertising campaigns, to see that this does not come about!

Microorganism Pre-Requisites

We know that people fed from synthetically-treated soils, generation after generation, will quickly develop chronic degenerative diseases. Fundamentally, the microorganisms, which are more important for the liberation of nutrients into plants than the fertilizers, cannot grow very well in synthetically-prepared soils for two reasons. First, synthetically-raised soils produce plants with weakened immunity and resistance to pestilence. Thus, pesticides are used that retard the soil's population of micro-organisms. Secondly, synthetically-treated soils are not being replenished with the full complement of trace minerals in the exact ratio and balance as occurs in nature. This balance ratio is not replicatable by the hand of man. Thus, these soils cannot feed the demands of a healthy microoganism population.

Examples of chronic degenerative diseases being caused by synthetically-raised soils (which must necessarily eventually lead to trace mineral deficiencies) in man are heart disease, cancer and growth problems, among others.[22]

The addition of cheap trace minerals to synthetically-complete fertilizer would appear to once again take care of the problem, when in actuality it cannot. Additionally, this in no way corrects the cause of the problem, which seems to me to surround the issue of the microorganisms. If the microorganisms can't live in these treated soils, what other degenerative processes await us, currently unknown due to the loss of this critical and essential link in our food chain? Indeed, how can we expect to thrive on depleted soil ourselves if it cannot even support the measly growth of single-celled organisms?

Hence, synthetic fertilizers destroy an essential continuum of *The Genesis Effect*, resulting in an insidious rise in the *hidden hunger* phenomenon. After witnessing first hand just what rejuvenating properties occurred in Findhorn where they used organic\bio-dynamic techniques exclusively, you couldn't convince me otherwise.

Blue Green Algae as a Rich Nutritional Source

What is the relationship between organic/bio-dynamically raised produce as discussed above and blue green algae? Wild-grown blue green algae is perhaps the only remaining stronghold of the optimal expression of mother nature's hidden manna. One species grows in a supersaturated stronghold of ancient rock dust in Upper Klamath Lake. As a result, one can obtain all of the vital nutrients and enzymes needed for proper body function from blue green algae. This is important because it is very difficult to get them from our current food system.

Taste as an Indicator of Nutrient-Rich Produce

Remember that one simple test you can use as an indicator for produce high in *Genesis Effectors* is to taste how "sweet" the produce is. This is easily discernable with carrots. But other vegetables are also easily taste tested. Also, what may seem sweet to you, may not be nearly what the vegetable might taste like if it were raised optimally using organic/bio-dynamic methods. Try to buy some freshly-harvested produce from your local health food grocery that is certified organic produce. You may notice a difference, but remember that in Scotland, even the organic produce could not compare to the produce grown in Findhorn's organic/bio-dynamic gardens.

With blue green algae, the taste will not be sweet, but strong, nutty and slightly pungent. This taste is also medicinal in quality, lending insight into its cleansing nature. The **aliveness** is almost discernable as one sucks on a tablet, something for which one can soon acquire a taste if given the chance.

Other vegetarian dietary sources of high quality *Genesis Effectors* are cold processed alfalfa, wheat grass tablets, seaweed, bee pollen, sprouts, raw vegetable juices and royal jelly, just to name a few. In general, *Genesis Effectors* **originate only from raw food sources!** There are both vegetarian and non-vegetarian sources. Remember that *Genesis Effectors* lead to regeneration within the body.

I want to point out that all substances known to induce regeneration come from raw food sources. I know of no man-made, artificially concocted or processed chemical that can induce regeneration within a living creature. Regeneration agents, therefore, lie exclusively within the domain of raw food factors.

Genesis Quicknotes

1. Synthetic fertilizers cause nutrient deficiencies within soils. Organic\bio-dynamic techniques enhance soil nutrient bio-availability by: (*a*) providing balanced nutrients; and (*b*) achieving optimal growth of soil microorganisms. Plants grown in these soils exhibit high natural resistance to infestations.
2. Synthetic fertilizers induce gradually increasing demands for the use of pesticides and insecticides. This results in a catastrophic die-off of the soil's essential microorganisms.
3. Synthetic fertilizers unequivocally induce degeneration into our food chain.
4. Newer studies point out that organic methods not only yield twice as nutritious produce, but also equal the yields of chemical methods. In fact, there are many reports that organic methods dramatically build crop immunity.

CHAPTER THREE

THE DISCOVERY OF *THE GENESIS EFFECT*

To continue any longer as blind consumers of life, without learning to be visionary restorers of life, will likely insure an end to both opportunities.

Donald A. Weaver

Introduction

Dr. Alexis Carrel (1873–1944) was a French-American surgeon and scientist. He was born in France and received his medical degree in 1900 from the University of Lyon. In 1906, as a promising young scientist, he joined the Rockefeller Institute for Medical Research in New York. He received the Nobel Prize in medicine in 1912 for work he had begun where he successfully sewed blood vessels together. This was a necessary precursor to the first organ transplants being performed.

In 1912, Carrel started perhaps the most inspiring experiment ever conceived in the principles of regeneration. Because of the immensity of this work, most of the regeneration researchers since his time have credited him as being the "Father" of regeneration therapy. I shall quote from Novak on this experiment:

> In January 1912, Carrel succeeded in transplanting heart tissue from a chick into an in vitro (outside the body) culture. He maintained the culture in an alive, although primitive, state for 38 years. Carrel was especially careful in providing the correct medium for the tissues. He added embryonic chick juice or some

> other liquid nutrient to the original artificial medium. He also bathed the tissue in fresh nutrient and discarded the used medium to ensure the removal of waste products.[23] (parentheses added)

Consider this incredible feat:

1. Carrel kept chicken tissue alive and perfectly healthy for well over thirty years beyond what it was expected to normally live.
2. No invading organisms infected these tissues during the entire course of the thirty-eight year experiment, even though exposed to unsterile open air.

Carrel described his technique a little further: "The first technique by which cells could be kept indefinitely in a condition of constant activity consisted in removing the tissue fragment frequently from its medium, washing it in Ringer solution, and transferring it to a fresh medium."[24] This frequent changing was done every forty-eight hours, not enough time for the nutrients to have been consumed by the tissue to any significant degree. Therefore, the frequent changing was more related to assuring for the removal of **all** waste products.[25] Since it is not possible to simply "wash" invading organisms away in such a manner, how did the tissue, without the benefit of an intact immune system, stay disease free? Could it be that the careful and thorough removal of wastes gave no foothold for invaders to multiply?

In this and subsequent volumes of *The Genesis Effect*, we will discuss the significance of Carrel's experiment, including the five outstanding and integral parts of his experiment as follows: (*a*) very special nutrients required; (*b*) optimal oxygen saturation; (*c*) detoxification; (*d*) longevity; and (*e*) pH. We will demonstrate that blue green algae provides the essential factors for all five of the above criteria. (The first three will be discussed in this volume, and the last two will be discussed in future *Genesis Effect* volumes.)

Genesis Effectors: The Special Nutrients Required —

At first, the main activating substance that kept tissues alive outside of the body were referred to as "embryonic growth-promoting substances" or trephones.[26,27,28] Carrel noted that "trephones" or embryonic growth-promoting substances were destroyed upon heating above physiological (body) temperatures.[29]

As you can already foresee, trephones are one and the same as *Genesis Effectors*, and blue green algae is extraordinarily high in these. From this point on, I shall refer to these embryonic growth-promoting substances, trephones, simply as *Genesis Effectors*.

At the time of these discoveries, there were some important conclusions drawn regarding exactly what these *Genesis Effectors* were. Some thought they were, in part, the newly-discovered group of known and unknown vitamins. Others thought they were, in part, the newly-discovered group of known and unknown hormones. However, some important facts about *Genesis Effectors* were that they could only be found nutritionally complete in all their varieties in very young tissues (with the only exceptions being adult white blood cells and adult glandular tissue).

Blue green algae is a perpetual embryonic tissue source. This is due to its incredibly high natural reproductive rate under optimal conditions. Such correlations occur only in the wild state, whenever and wherever there is an abundance of available nutrients for algae growth. These "algal nutrients" are then abundantly converted into *Genesis Effectors* within the algae.

Furthermore, note the following: although cells deprived of *Genesis Effectors* can divide and reproduce, they **do not grow** without nutrient initiators. Rather, they will merely split in half what cell mass they have. As a result, there will be twice as many cells or more, but no net change in cell volume as a whole. When you introduce *Genesis Effectors* such as embryonic juices, rapid cell growth occurs, drastically increasing the size of each and every cell.

Could stunted growth be a result of a lack of raw foods in our children's and animals' diets? And what about adults? Could dietary deficiencies of raw food factors contribute to poor cell renewal, and hence accelerated aging? Embryonic *Genesis Effectors* must be preserved for use into adulthood when needed. Hence, the glands of the human body attempt to develop reserves of these factors. However, today, most people living in industrialized nations are bankrupt in this regard. This is where concentrated sources of *Genesis Effectors* can play such a vital role. Blue green algae can accomplish this task well.

It has been theorized that "mother" (progenitor) embryonic growth-promoting *Genesis Effectors* give rise to the individual adult organs that are endowed with a limited reserve of *Genesis Effectors*. Hence, nature supplies us with continued access to these vital essences. Blue green algae helps free these essences back up within the cells for the body to use in regeneration. On a much grander scale, blue green algae acts as the entire planet's primordial progenitor, since it is both the first-known life form on the planet, and the initial life form from which the entire planet's food chain arises.

Blue green algae's continuous evolutionary process keeps it current to present-day planetary conditions. Blue green algae seeks to adapt to current global stresses and environmental antagonists, thus maintaining regenerative forces applicable to the twenty-first century. In other words, blue green algae serves as an adaptogenic regenerative agent, in step with the pulse of global living.

Adult Gate Control Mechanisms for Regeneration

As stated, adult tissues that display a gateway mechanism (that is, cells that harness and direct the regenerative efforts of *Genesis Effects*), are the white blood cells and the endocrine organs. Collectively, they are known as the tissues that allow *The Genesis Effect* to acquire "Threshold Potential." As Carrel stated:

> The presence, in the embryonic tissue juices and in glandular and leukocytic (white blood cell) extracts from adult animals, of principles increasing the rate of multiplication of fibroblasts and epithelial cells (young connective tissues) suggest the possibility that these substances are secreted by leukocytes and endocrine glands.[30] (parentheses added)

Genesis Effectors induce an increase in metabolic functions, growth, replication, and thus regeneration as well. This is a "generalized" effect. *Genesis Effectors* do not by themselves selectively work on one tissue and not the other. *Genesis Effectors* work on all of the tissues they reach. Thus, there must be adult tissues that have inherited the regenerative responsibility to surgically focus regeneration wherever needed. It has been proposed that not only do our own internal white blood cells and glands retain *Genesis Effectors*, but also that these two separate systems interface to work in concert to surgically focus regenerative actions. Recall that the adult white blood cells and endocrine glands were found by Carrel to display all of the same characteristics as embryonic juices.

What this means is that the human system, by its inherent design, when functioning and properly nourished, should be able to avoid chronic degenerative diseases. Blue green algae provides for restoring all facets of the **gate-keeper** mechanisms, as found within the immune and endocrine systems, wherever and whenever these adult systems become damaged, depleted or malfunctioning.

Genesis Nutrient Initiators

To better illustrate blue green algae's ability in maintaining, reestablishing or regenerating all aspects of *The Genesis Effect*, we must correlate other studies on this subject.

The liver, one major storage site of nutrient initiators, has remarkable regenerative properties, retaining many of the abilities of the embryonic *Genesis Effectors* into adulthood.[31,32] We know that hormones,[a] colloidal minerals and vitamins alone cannot account for **all** of the amazing properties of the embryonic *Genesis Effectors* found within the liver.

Recall that embryonic *Genesis Effectors* are destroyed by heat. This property alone makes it automatically next to impossible to analyze such factors. Why? Because when scientists attempt to analyze something, they have to isolate it by heating it, boiling it, distilling it with solvents, precipitating it out with potent chemicals and/or bombarding it with electrons or x-ray radiation to photograph it, rip it apart with special filters, etc. All of these processes, to various degrees, destroy *Genesis Effectors*. In the end, only incomplete information can be gleaned with the current technologies available. Thus, it is important to avoid making any steadfast conclusions or assumptions regarding *Genesis Effectors* since such conclusions would not be based on the entire picture.

Although I have stated that we might not ever be able to analyze and identify *Genesis Effectors*, some information is, nevertheless, available. We can partially address several special substances currently known to possess potent embryonic rejuvenating effects.

Structural *Genesis Effectors*

Epidermal growth factor (EGF)—This first-line commander is a polypeptide, similar in composition to hormones, that stimulates the major activity of growth and reproduction in a wide variety of cells.[33] EGF stimulates the cell membrane, which in turn activates two key enzymes that regulate growth. Blue green algae contains at least one substance that has demonstrated strong affinity to EGF (i.e., RNA).[34]

Ribonucleic Acid

Raw RNA—or ribonucleic acid is another potent *Genesis Effector*. Formerly, it was **erroneously** thought that nutritional RNA sources only

---

[a]*Genesis Effectors* (in this case, embryonic) also include hormone effects, but they are not limited to simple hormone effects in that *Genesis Effectors* are nutritional substances that add nitrogenous (i.e., protein) growth to cells, and are used up in this growth (therefore serving as substrate material); whereas, hormones act fundamentally as catalysts that accelerate metabolism without themselves being used up as substrate materials.

supplied simple nutritional replacement parts for the growth factors.[35,36] But it is now well known that they do indeed actually induce much more than this. Taken orally, raw nutritional RNA induces sophisticated *Genesis Effects* such as reproduction and unscheduled growth activation in tissues. Since nutritional sources of RNA come in slightly different forms, some forms may be more stable to digestive actions than others. As expected, the forms most likely to survive unscathed during digestion (and thus be available for human tissue regeneration efforts) are the forms most powerful in their *Genesis Effects*.

Additionally, it was thought that there were no receptor sites on the cell membrane to bring nutritional sources of RNA intact into the cell, if indeed nutritional RNA could survive digestion. It was concluded, therefore, that there is no value in RNA other than simple nutritional effects. Although there may be no current known mechanism for dietary RNA to enter into a cell intact, it does not mean there isn't one.

Nutritional RNA may be quite small, thus it might simply "slip" in by way of a membrane pore. Whatever the reason, there is ample evidence today that nutritional RNA does get into both your body as well as your cells whole, **or at least the active residues do which still retain all of the growth-promoting properties**.[37,38,39,40,41] Additionally, nutritional RNA is a strong inducer of rapid protein growth inside specific cell tissues.

Wild-grown, properly-preserved food grade blue green algae from Upper Klamath Lake may be the richest, most therapeutic, **least toxic and most cost-effective** known source of nutritionally-available RNA.

Optimal Oxygen Saturation and RNA

Robbins and Cotran stated in their book, *Pathological Basis of Disease*, that: "Lack of cellular oxygen is probably the most common cause of cell injury and may also be the ultimate mechanism of damage."[42] In fact, a nurse who specializes in medical research said to the author of the last footnote reference, that, "It's so simple. I don't know why I never thought of it. When we're working with cell cultures in the lab, if we want the cells to mutate, we turn down the oxygen. To stop them, we turn the oxygen back up."[43]

Basically, when cells are engaged in metabolism, they are "burning" fuels delivered from your diet. This "burning" uses up oxygen so that energy can be recovered from the metabolic breakdown of foodstuffs. Interestingly, this recovery of energy depends ultimately upon oxygen combining with hydrogen in the final step, something which largely depends upon Coenzyme Q10 (CoQ10). It is now well known that RNA spares and optimizes oxygen utilization within the body, especially when combined with CoQ10. CoQ10 is manufactured in our body from raw food intake. As we age, this vital enhancer to *Genesis Effectors* significantly diminishes. Therefore, even people eating high raw food diets should consider taking supplemental CoQ10 daily to augment blue green algae's fullest potential.

Cooked Foods Lose Oxygen

There is another type of "burning" that also results in using up oxygen. When you cook your foods, you are expelling the oxygen present in the raw food. This causes cooked food to be more acid forming in your system. This is especially true of cooked proteins and fats.[44] *Genesis Effectors* are laced with oxygen, and are destroyed when the food is partly cooked because the oxygen is thrown off. Additionally, processed foods contain large amounts of food additives, hydrogenated fats and preservatives, all of which rob us of even more oxygen. So you can see that we not only breathe oxygen in through our lungs, but we also were designed to eat it in our raw foods as well. In fact, Carrel demonstrated that when embryonic tissue juices are heated up, the *Genesis Effectors* are not only destroyed, but *growth-inhibiting* substances result.[45,46]

Pre-Requisites of Regeneration

This volume will address blue green algae's impact on humans and animals. It is important to know from the outset the three *basic* ingredients required for inducing successful regenerative phenomena. **First, there must be detoxification. Without detoxification, no regenerative therapeusis can occur. Second, there must be optimal saturation and utilization of oxygen for all metabolic pathways. Third, without raw embryonic-sourced nutrients, there can be no true unscheduled regenerative phenomena.** And most importantly about these three, each must be understood *mentally* and *physically*. To do anything less will take no one to the final goal–true regeneration! Each depends upon the other, but they are also sequentially dependent as well. When these criteria are successfully brought together, it is called *The Genesis Effect*. This volume will describe the various qualities of blue green algae that single-handedly emulate these essential regenerative pre-requisites.

Unraveling Regeneration

The Genesis Effect prevails throughout all of nature, both on a global scale as well as within all of earth's creatures. Global cycles of regeneration and degeneration are nowhere better described than by Hamaker.[47]

On a macrocosmic scale, *Genesis Effects*, as they occur in our bodies, are linked to *Global Genesis Effects*. Without soil renewal, which necessarily must include putting *all* mineral elements back into the soil, all life on this planet gradually degenerates. On a microcosmic scale, *Genesis Effects* occur within our structure and function. One way to understand how this occurs in higher life forms is to first define degeneration. There are two basic types of degeneration—**structural and functional**.

Regeneration/Degeneration

Structural degeneration occurs when both pattern-making molecules and boundary-separating molecules start to break down. The most important of these are: (*a*) the nucleic acids, which contain your genetic code; and (*b*) cell membranes that separate the machines of cellular mechanics. Cell membranes wear down slowly as aging sets in, altering critically-precise messages to the nucleic acids and causing a cascade effect (an acceleration to the aging process).[48]

Functional degeneration occurs when the **functional** processes in the body begin to break down. The most important inducer of this type of degeneration is the over-accumulation of inorganic and organic waste acids in the tissues. These acidic wastes consume or block oxygen utilization. This leads to a cascade event that systemically inspires further degenerative processes. So, how do we begin regeneration? For most people, the answer is simple:

1. Remove poisonous wastes from within the body.
2. Stop or minimize their reintroduction or production within the body.
3. Enhance oxygen delivery to all body tissues.
4. Consume raw embryonic nutrients such as those found abundantly in properly-preserved wild-grown blue green algae.

Blue green algae's high RNA, chlorophyll and antioxidant content satisfies all of the above requirements.

Reversing Structural Degeneration

Damaged structural pattern-making molecules in our body must be replaced. Blue green algae has replacements for such molecules. Additionally, blue green algae, because of its high antioxidant content, protects and prevents future breakdown to this regenerated structural pattern.

Reversing Functional Degeneration

Blue green algae is rich in nutritional precursors for the functional pattern-making substances. For example, blue green algae is super rich in neurotransmitter precursors, enzymes, chlorophyll, all known vitamins (except vitamin D which we get from sunlight) and minerals, and a plethora of associated food substances and other unknown factors that bring about ideal conditions suitable for regeneration.

Genesis Quicknotes

1. Nobel laureate Alexis Carrel offered the first sound experiment that suggests tissue immortality.
2. The three pre-requisites to regeneration are: (*a*) embryonic-like nutrients; (*b*) optimal oxygen saturation; and (*c*) detoxification.
3. Civilized living has overwhelmed intrinsic regenerative powers and mechanisms.
4. Food grade blue green algae as a solo tool, supplies in abundance many of the most powerful regenerative substances yet discovered, without any known toxicity.
5. Blue green algae, due to its inherent composition, may be without equal in optimizing tissue oxygen saturation and utilization.

Chapter Four

Blue Green Algae and Repair: *Genesis Effector* Supreme

We cannot retard senescence or reverse its direction, unless we know the nature of the mechanisms which are the (basis of longevity).

Alexis Carrel

Blue Green Algae and RNA: Justification for Its Oral Use in Regeneration Therapy[a]

Up until quite recently, major medical texts taught that protein is completely broken down into simple amino acids in the digestive tract. Recent studies clearly refute this idea. However, there is still widespread bias against survival of RNA as well as protein in the digestive process. Two important biochemistry texts have carefully covered the digestive pathway of nucleic acids, and they both stated that nucleic acids are completely broken down into non-biologically-active substances other than for some rudimentary nutritional effects.[49,50] However, these texts are over twenty years old, and an update is long overdue.

[a]Blue green algae is 4% RNA and nucleic acid – perhaps the richest source known.

Recall that there are receptor sites looking for specific minerals to transport across the gut wall. RNA is often found attached to minerals such as copper and magnesium, so this copper or magnesium complex might facilitate intact RNA absorption.

Additionally, nucleic acids are full of energy-rich components that the body needs. One scientist, Durkacz felt that nucleic acids can be transported through the gut using the same mechanisms as viruses. Recently, studies performed at the National Institute of Health discovered that a special substance called "lambda protein" wraps around nucleic acids, protecting them from digestive breakdown.[51]

Another explanation why biologically-active fractions of RNA may survive trans-gut absorption is that some RNAs are a bit more stable than others, as already mentioned. You should note that there are three types of RNA—r-RNA, t-RNA and m-RNA. All three groups fall within the size ranges shown in Table 4.1 (see page 31) which are conducive to intact intestinal absorption. But r-RNA is specifically shaped into a "helical-built" molecule, and is found in active fractions attached to a protein. This "helical-built" property lends excellent stability helping to resist possible enzymatic breakdown.

In the laboratory, when this RNA-protein complex was saturated in a test tube for a continuous twenty-four-hour period with a **proteolytic** enzyme, all activity was lost.[52] However, that was under laboratory conditions. What about conditions found under normal circumstances within our bodies? We know that up to 46% of some dietary protein survives digestive breakdown. Thus, we have good reason to believe that significant amounts of biologically-active RNA reach our tissues after eating foods containing RNA.

Trans-Gut Absorption

Recently, many elegant studies reported in the peer-reviewed literature have confirmed beyond doubt that intact absorption of biologically-active substances does take place. Gardner has reviewed many of these reports and stated:

> However, one special aspect that has been largely neglected hitherto is the possibility that peptides produced during digestion in vivo of a protein meal may enter the circulation in intact form and that they may thus reach peripheral tissues where they could exert biological activities . . . There is now a substantial body of evidence albeit not widely known that significant quantities of larger molecules including peptides and even proteins can cross the intestine.[53]

Gardner gave an excellent review of the best understood mechanisms involved with transporting intact biologically-active substances (i.e., *Genesis Effectors*) across the intestinal membranes and into general circulation:[54]

1. Pores that are obvious trans-cellular passageways.
2. Fat or lipid regions that may provide easy trans-cellular passageways.
3. Area-mediated transport by specific trans-cellular mechanisms.
4. Pinocytosis through normal but specialized epithelial cells.[55,56] The best example of this process deals with the absorption of antibodies. Bradford described it as follows:

> The surfaces of these intestinal cells then depress (invaginate) into the cytoplasm and pinch off small membrane-bound sacs of bubbles (phagosomes). On the inner lining of the phagosomes are found the IgG antibodies bound to their receptors in the membrane. The solution trapped within the phagosomes contains solubilized food destined for digestion.[57,58]
>
> The phagosomes then merge within the cytoplasm, with other membrane-bound sacs, the lysosomes, which contain digestive enzymes. The resulting sacs (phagolysosomes) are pulled through the cytoplasm by microtubules to eventually contact and merge with the outer (plasma) membrane of the cell. Here the products of digestion are emptied into the extracellular space and eventually enter the circulatory system.
>
> While in the phagolysosome, soluble food is subjected to the action of digestive enzymes (including proteolytic) originally present in the lysosome. The IgG immunoglobulin bound to its receptor site on the inner surface is also subjected to the same proteolytic action. About 40% of the IgG molecules survive digestion for two reasons. One is the protection afforded by the receptor site itself. The second is related to the structure or manner in which the protein strands themselves are folded within the molecule (the conformation). The protein structure of IgG is that of a "beta-barrel" or "pleated sheet" as contrasted with the "alpha-helix."[59]

   These structures represent two basic conformations of protein found in nature. The beta-barrel form of protein is more resistant to proteolytic action than the other conformation.

   The process of antibody absorption described above diminishes as the intestinal lining matures. **However, it has been shown that significant vestiges of it persist into adult life**.

5. Pinocytosis through specialized "M" cells. These overlie the Peyer's patches and permit lymphocytes (a type of white blood cell) to come in very close to the intestinal lumen. The possibility has been advanced that this lymphocyte-mediated transport system allows antigens to be quickly identified for the appropriate immune

response. In a totally natural event, white blood cells are able to collect *Genesis Effectors* at this point.[b] The white blood cells are the best candidates for mediating the "target tissue specificity" transport mechanism[60,61] to induce regeneration with precision.

6. And finally, there are special para-cellular passageways extending through tight junctions and extrusion zones similar to the pores listed in 1. above.

The question now becomes not if this happens, but rather how much quantity can pass on through, and how large can these biologically-active substances be?

How Much Gets Through?

Earlier studies revealed that only 2% of very large proteins would be able to pass through the intestinal walls whole and into the blood stream.[15,62] Then a little later, with electron microscopy and sophisticated radioactive tagging procedures, Hemmings showed that no less than 5% of whole proteins are absorbed,[63] and up to a high of 46% of whole proteins could be absorbed! In fact, more than 20% of what is absorbed intact completely retains all of its original characteristics. Radioactive tagging procedures are now done rather routinely, and it is generally agreed that when this radioactive substance is found at some internal location, because of the way in which "tagging" is set up, the original characteristics of the substance remain intact.[64] Studies performed on RNA have yielded similar results, meaning that the biologically-active forms of RNA did arrive at certain target tissues. In fact, Hemmings asserted that biologically-active substances such as protein:

> . . . might be termed distributed digestion, occurring in all body cells rather than in the gut lumen. . . . Tentatively, it is put forward that there is a universal necessity for body cells to be "permeable" to proteins of many types and that there is a constant traffic of protein molecules into all body cells.[19]

What Sizes of Biologically-Active Substances?

Colloidal bound minerals, vitamins, hormones and other nutrients easily pass through the intestines and into the blood stream intact. Generally, these substances are around the same size ranges. The unit of measure for these substances is Molecular Weight (MW). Size is not the sole criterion

---

[b]In another volume by this author, *Glandular Therapy*, vol. 1, a scientific text written for health professionals, a discussion is presented.

Table 4.1 Sizes of Biologically-Active Substances Known to be Absorbed Intact

Biologically-Active Substance	Molecular Weight (MW)
Vitamins—U.S.P. Grade	123.1 to 1,357[65]
Vitamins—Food Incorporated[66,67]	1,000 to 5,000[a]
Insulin	6,000[68]
Lysozyme	14,300[69,70]
Chymotrypsin[71,72] (an Enzyme)	23,000[73]
Lipase	30,000[30,31]
Horseradish Peroxidase[74]	40,000[75]
RNA (an Active Fraction)[76]	1,000 to 5,000
RNA[77,78]	23,000 to over 1,000,000[a,79]
Immunoglobulin IgG[80]	150,000[81]
Ferritin	750,000[82]

[a]These figures are approximations based on the best available data

as to whether or not a substance can gain access to the blood stream from the gut, but it is significant enough that we may speculate here. In fact, paradoxically, some larger-sized molecules may actually pass through faster than smaller ones.[83] Table 4.1 shows the known size ranges of biologically-active substances that have been documented to pass through the gut intact.

For future reference, let's arbitrarily set the 50,000 MW class as a realistic range for any nutrient to pass through intact to some degree and in biologically-significant enough quantities to affect beneficial functions. Your body is endowed with mechanisms for special transport of select nutrients. It is no mistake that we have these transport mechanisms to acquire fresh supplies of *Genesis Effectors* from our diets.

Finally, remember that proteins in general are comprised of amino acid chains, and some amino acid sequences are very stable under enzymatic action, while others are not. However, Gardner stated that since **all** of the different varieties of amino acid sequences are represented in his survival studies, there must be other hidden mechanisms that prevent enzymatic breakdown.[84] In summary, your body is well equipped to absorb all of the health benefits that lie within blue green algae, whether they are amino acid based or not!

RNA: Nucleic Acid Composition

A brief review on RNA and DNA is in order here. Nucleic acids (DNA and RNA) are composed of simple structures called free "**bases.**" These bases are the fundamental building blocks of our bodies. From these fundamental building blocks, bases become a little more complex. As they add on energy-rich molecules, they become **nucleosides**.

Nucleosides can donate their energy storehouse to more complex molecules such as DNA or RNA. In fact, nucleosides evolve to the next level by adding on more and more of these energy-rich molecules. The most famous evolved nucleoside is **adenosine tri-phosphate**, which is the **major currency your body uses for energy**. Adenosine tri-phosphate actually qualifies as a "nucleotide." Nucleosides are a family of crucial molecules dominated by four main types. Just think of these as the major currencies of the world (i.e., German Mark, British Pound, U.S. Dollar, Swiss Franc, etc.) with the strongest currency being adenosine tri-phosphate, which would be analogous to gold. Then, nucleosides are made a little more complex and energy rich with the addition of one more molecule. Now, they are called **nucleotides. Can you now see why many people consuming optimal amounts of wild-grown blue green algae report a drastic increase in their energy levels?**

Nucleotides, as **coenzymes**, supercharge normal metabolism. *Active* vitamins assist by *optimizing* normal metabolism. Remember the various members of the **B**-Complex family of vitamins? These serve as coenzymes in their active forms. So, think of nucleotides as the **N**-Complex family of coenzymes. The only major difference is that these nucleotides are **energy-rich** molecules because they contain nucleosides.[c] This energy is what your body uses for all metabolic, growth, regenerative and reproducing processes. So you can now see how important it is to have enough energy available in the form of these building blocks. For this reason, I call **nucleotides** the ready, willing and able molecules of energy for the body. Additionally, "modified" nucleotides are the major intermediaries of most all hormone *Genesis Effect* actions.

RNA injections are well known to increase the growth and protein synthesis of tissues from 50% to 100%.[85,86] Thus, your food sources and

---

[c]Nucleotides and nucleosides are, however, not regarded as dietary essentials like vitamins. But because they harness so much energy, you should begin to see that they are essential to your diet if you want to prevent chronic degenerative disease.

your own recycled tissue sources of RNA could, therefore, specifically regenerate your damaged tissues.

DNA is virtually the same in every cell and tissue of your body. I doubt that it could display any real *Genesis Effect* for this reason. Cameron conducted studies on both DNA and RNA to find out the differences between them. His findings were that DNA displayed no positive effects in his study, while RNA absolutely did.[87,88]

As far back as 1951, Newman conducted feeding experiments with both the individual and separate components, free bases, to the RNA molecule. He erroneously concluded that the free bases of RNA displayed no target tissue regeneration specificity (TTS) or any other action of special biological importance, except simple nutritional effects. However, he found that when the nucleotides were fed, they were incorporated intact into cellular nucleic acids, a critical feature of repair.[89] But what about larger and more biologically-activating fractions of the nucleic acids? Or better yet, can **whole** intact RNA be incorporated to repair a recipient's damaged tissues?

As previously mentioned, historical objection to this idea is replete with sound logic (i.e., the fact that there are no known receptor sites on any cells for the RNA to be transported inside.[d])

Recall that RNA can be attached to minerals. So can the nucleosides and nucleotides, which especially like calcium, magnesium and copper ions.[10] RNA, although quite large, may make its entrance by tagging along with the enhanced mineral transport mechanism. Also, we recently mentioned the discovered "lambda" protein that can protect dietary nucleic acid digestive breakdown, and remember Durkacz's suggestion about a nucleic acid trans-cellular transport mechanism. Recently, Rothschild has shown that some enzymes (and plausibly RNA) may become wrapped up into a liposome during digestive processes, allowing for intact absorption.[90]

Perhaps the most definitive study done to date on the fate of orally-ingested nucleic acids comes from Greife and Molnar. They used sophisticated radioactive tagging studies that clearly showed how previous investigators all failed to see RNA's regenerative qualities (such as Newman who missed two key discoveries). First, in contrast to Newman, Greife and Molnar were able to show that free bases **are** incorporated

---

[d]However, it is known that if you place tissue into a dish and destroy the functioning RNA via enzymes, you can re-instill fully-functional RNA fractions by adding outside RNA into the dish. How does this freshly-added RNA get into the cell to promote normal RNA functions (i.e., protein synthesis) if there is no cell membrane mechanism to bring it in? Obviously, there is indeed some unknown mechanism.

whole in a specific cellular fashion within cells. More importantly, they also found that nucleic acids, when fed orally, displayed a strong specific regenerative effect. They labeled nucleic acids with a radioactive tag that would reveal that the intact nucleic acid did enter into specific tissues **if** the radioactivity could be traced to the tissue. Remember that it is generally agreed that radioactive tagging is hooked onto a molecule in such a way that it remains with the biologically-active fraction. Therefore, no radioactivity should show up any place other than with the intact biologically-active molecule.

The results? Twenty-four hours after dosing orally, **19.7%** of the nucleic acids could be found in target tissue specific areas.[e] **Twenty-one days after oral administration**, the nucleic acids were actually found in even higher concentrations than the free bases that were given by injection (5% versus 2%, respectively)! The target tissues in this study were the liver and kidneys. Therefore, oral feeding was superior to the injection route for inducing unscheduled growth activity in these locations.[91]

In summary, Frank listed the six most important attributes of RNA Therapy:[92]

1. Anti-anoxia/Hypoxia effect.
2. Adaptogenic effect over severe temperature fluctuations.
3. Anti-tumor growth effect.[93]
4. Anti-aging effect.[94]
5. Anti-viral effect.[95,96,97]
6. Ergogenic agent, meaning it enhances endurance and overall energy levels.

To this list we must add the following:

7. Anti-radiation effects (e.g., x-ray protection).[98,99,100,101]
8. Wound-healing accelerator.[102,103,104,105]
9. **Partly reverses inherited and congenital genetic damage** (i.e., mental retardation, Down's Syndrome, etc.)[106,107]

By thoroughly detoxifying our tissues with a concise program, we can rid ourselves of most water-soluble toxins within three to six weeks, and most fat-soluble toxins within three to six months. After detoxification, provided that adequate *Genesis Effectors* are bio-available within our repair mechanisms, we suggest that specific tissue regeneration will proceed at the known tissue turnover rates listed below:

---

[e]Target tissue specific areas means that when one ingests raw heart tissue, regenerative effects are induced to a greater extent in the recipient's heart than in any other organ or tissue.

Stomach lining – replaces itself once per hour;
Intestinal lining – replaces itself every four days;
Heart tissue – replaces itself once every ninety-two days;
Blood – replaces itself once every one-hundred twenty days;
Most organs – replace their inherent tissues once every five to six months;
Bone – replaces its cells once every three and one-half years.

In conclusion, if we are diligent and persistent (with the exception of the nerve cells), we might actually achieve *total* body renewal within this three and one-half year time frame. Therefore, once convinced, a wise person will remain steadfast on a regenerative program, knowing that he/she is bound to be successful in the long run. Remember, a journey of a thousand miles begins with the first step. Chapter Five is your first step!

Genesis Quicknotes

1. Science confirms nutritional RNA's survival and delivery into consumer's cells.
2. Science confirms gateway **mechanisms** that logically allow for nutritional RNA transport across the gut lining.
3. Science confirms that significant **quantities** of nutritional RNA can feasibly pass through the gut lining unscathed.
4. Science confirms the existence of logical nutritional RNA **delivery** mechanisms delivering RNA into precisely-focused tissue locations.
5. Science confirms intense **regenerative phenomena** associated with nutritional RNA intake.

CHAPTER FIVE

WHY WILD BLUE GREEN ALGAE REGENERATES

The time may come. . . when the pressure of population on food supplies will justify mass cultivation of blue green algae. . . .

Scientific American, 1966

Most of the food we eat comes from weakened soil.

Firman E. Bear, 1962[108]

Blue green algae is not a vitamin supplement, mineral supplement, herb or drug. *Blue green algae is the most nutrient rich "whole" food on the planet.* As such, it will fill in many of the "blanks" and deficiencies now unavoidable due to our daily diet. Since most of today's modern processed and grown foods are now nutritionally bankrupt, it is essential to replenish our lost internal nutrient reserves with this incredible whole food.

Is America's food supply nutritionally bankrupt? Bare in mind that there is a direct correlation between the nutrient content of food and the mineral content of the soil (see Chapter Two). In other words, foods acquire their nutrient value from the mineral wealth of the soil. In 1936, the U.S. Senate published document No. 264 which stated that America's soil was exhausted, leaving our foods nutritionally bankrupt.

> The alarming fact is that foods (fruits, vegetables and grains) now being raised on millions of acres of land that no longer contain enough of certain minerals are starving us – no matter how much of them we eat. No man of today can eat enough fruits and vegetables to supply his system with the minerals he requires for perfect health because his stomach isn't big enough to hold them.

> The truth is that our foods vary enormously in value, and some of them aren't worth eating as food. . . . Our physical well being is more directly dependent upon the minerals we take into our systems than upon calories or vitamins or upon the precise proportions of starch, protein or carbohydrates we consume.[109]

Author Dr. Michael Colgan, founder of the Colgan Institute of Nutritional Science of La Jolla, California, described this problem today: "Americans are malnourished. Even when we succeed at eating a balanced diet, our systems still suffer from the actual lack of nutrients in foods. For example, raw carrots differ widely in beta carotene content, often showing variances from 18,500 International Units (IU) to 70 IU per 3 1/2 oz. sample."[110]

Health is absolutely dependent upon complete nutrient availability and accessibility. This would suggest that most food grown in America has been incapable, since 1936, of instilling optimal health to its population; or even for that matter, sustaining health at all. The misconception is that if these facts were true, then it would be readily apparent to everyone. They are not apparent because the body's reserves are called upon, which are often (but not always) capable of buffering the true ***acute*** deficiency signs and symptoms, but not thc more insidious ***chronic*** ones. In effect, this pattern of going to the bank to make nutritional withdrawals, without ever making ***non-counterfeit*** deposits, destines us to nutritional bankruptcy. Thus, the obvious becomes well concealed, shifting Americans into a state of chronic illness, oblivious to where it all originated.

Today, nearly 50% of all Americans have some form of diagnosable chronic degenerative disease. Sixty million Americans have heart disease, with a frightening and escalating segment comprised of young adults and children. Over two-hundred thousand children suffer the torment of serious arthritis.[111] Cancer is also escalating at an alarming rate for children. In fact, cancer is the number one cause of death in the young. The National Cancer Institute tells us that 80% of all cancer arises out of the air we breath, the food we eat and the water we drink. According to the U.S. Surgeon General, 67% of all disease is diet linked. And adding insult to injury, the U.S. Public Health Department said in 1976 that only 1 and 1/2% of all Americans were actually healthy![112]

In 1948, America's food supply was analyzed and the results were published. However, since this report gave no reference to U.S. Senate document No. 264, the 1948 study had no frame of reference that it was, in the main, analyzing nutritionally-deficient foods. In 1963, there was a frightening drop in nutrient content from the previous 1948 food studies, as published in the Firman Bear report.[113] Today, this trend has become even worse. For example, an average bowl of spinach in the 1940's would have contained 154 mg of iron. By the late 1970's, the iron content of an average bowl of spinach dropped to 2.2 mg. Today, averages often fall below 1 milligram per bowl of spinach! This one mineral, iron, in this one crop, spinach, suggests that *other* crops suffer vast mineral losses as well. Why should spinach and iron be the *only* two involved in this trend?

Figure 5.1 shows the iron content of foods. The iron content of organically-grown and chemically-grown foods is shown for each vegetable tested in 1948. The number for 1963 shows the average iron content of both chemically- and organically-grown vegetables tested that year. Note that most 1963 averages are only slightly better or worse than the lowest values for 1948.

The figures for these crops are 30 years old. Where does that leave us today? The tragic conclusion is that there is no immunity from the devastating losses occurring throughout *ALL* of American-grown produce. Other specific examples of this trend were previously cited in Chapter Two.

Blue green algae, which is grown in a full spectrum mineral-rich environment, contains an optimal abundance of the nutrients that our bodies crave, thereby allowing us to escape this trend in today's food supply and extinguishing our *hidden hunger*.

How to Remedy Hidden Hunger

The miracle of algae began when it developed the ability to perform photosynthesis. Photosynthesis is the process whereby sunlight, water and minerals are converted into usable foodstuffs. Algae thus became the prerequisite for all higher life forms. Algae grow in virtually every niche and corner on our planet. Pigments of different colors found within algae are the basis for photosynthesis. Thus, they are classified into families by their color: blue green, green, red, brown, golden, etc. The sheer volume of algae is staggering. Taken as a whole, earth's algae population weighs 100 billion tons and stretches almost 8,000 miles across. By being the supreme regulator over the carbon dioxide and oxygen levels of our atmosphere, algae control the weather patterns. Algae stores, manages and controls 450 billion metric tons of carbon dioxide and 330 billion metric tons of oxygen annually. The result is that algae manufactures 90% of the oxygen in our air. In effect, the planetary impact and influence of algae upon life is unsurpassed.[114]

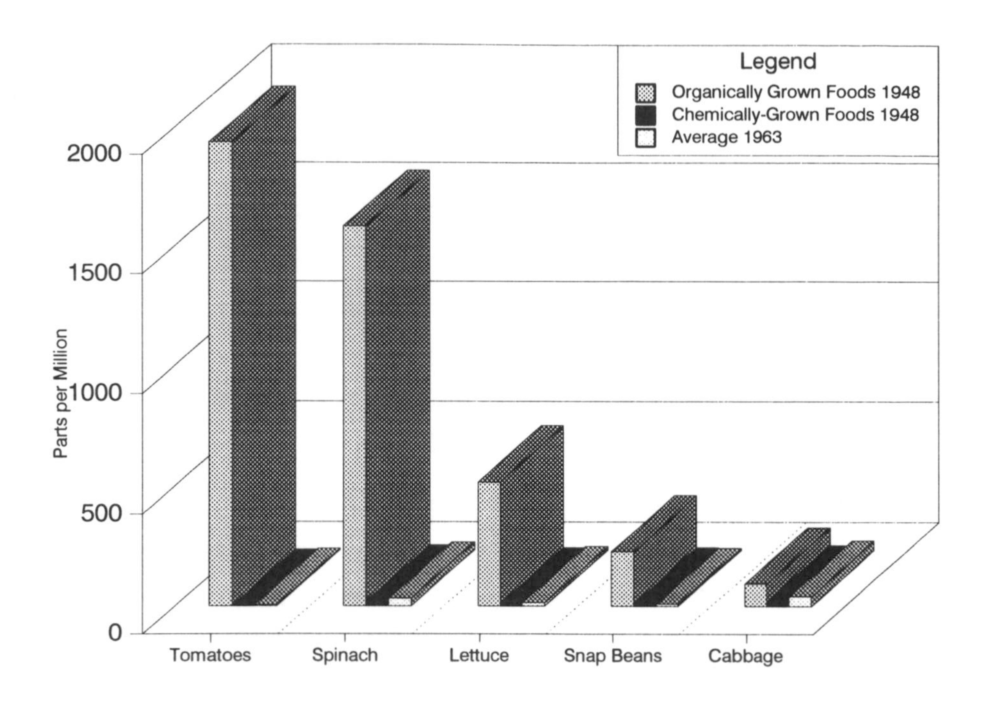

Figure 5.1 Comparison of the Iron Content of Organically- and Chemically-Grown Foods (1948 and 1963 Averages)

Source: *Firman E. Baer Report*, Rutgers University, 1984.

Since algae is the foundation to the entire food chain, algae enters into the diet (either directly or indirectly) of all known life. Taken as a whole, algae produce over five-hundred trillion tons of food annually on earth. This represents 80% of *all* the food produced worldwide. With over 30,000 different forms of algae, it is important to know which of these species best serves and is most compatible with the dietary needs of humans. Humans have been eating both fresh and salt water varieties of algae throughout history. These include many forms of seaweed and more recently, chlorella, a fresh water green algae. However, blue green algae (Cyanobacteria – "cyano" meaning blue) offers unique advantages to the human dietary that may far exceed all other foods available to mankind. Food grade *Spirulina* and *Aphanizomenon flos-aquae* are easy to digest, and are thus the two most important members of this family for humans.

From the U.S. Surgeon General we have the statement that, "Your choice of diet can influence your long-term health prospects more than any other action you might take." Now recall that 67% of all diseases are now known to be diet associated. Is it any wonder why blue green algae users have credited blue green algae as a miracle cure for such a wide spectrum of ailments? Can the reader appreciate how blue green algae might contribute to prevention and improvement of many (if not most) of the chronic degenerative diseases?

Blue green algae is best grown under ideal natural and wild conditions. For example, unpolluted *mineral-rich* seas or lake beds would be the ultimate source of food-grade blue green algae. Man-cultivated blue green algae is likely to be too profit motivated, again lending certain economical influences to cut corners, and thus compromising on blue green algae's optimal potential as a food. However, exceptions could occur if care was taken to augment artificially-raised blue green algae with rock dust and other sources of ultra fine, full-spectrum powdered trace minerals.

And most importantly, blue green algae should be harvested and immediately preserved so as to keep intact all of its inherent riches. Properly performed freeze-drying technologies now offer an excellent means to accomplish this. Thus, compliance with certain standards is necessary in the preparation of this nutritional whole food supplement to ensure effectiveness.[a]

Because it is a cholesterol-free protein source, it is also a perfect substitute for the large amount of animal fat in the typical American diet, and can thus help prevent heart disease. We shall see that blue green algae is an unexcelled source of minerals, vitamins, fats, amino acids, associated food substances and regenerative constituents.

Blue Green Algae As Food

In the words of Hippocrates, "Let thy food be thy medicine and thy medicine be thy food." We can now appreciate why end products of blue green algae must contain concentrated raw materials that demonstrate biological activity. These biologically-active raw food factors are composed of regenerative substances that will be discussed next.

---

[a]The reader is directed to a future publication on the subject of blue green algae preservation and processing from Genesis Communications, entitled *Blue Green Algae: Production and Safeguards*.

Protein Constituents of Blue Green Algae

BGA contains all 20 amino acids, including all of the essential amino acids. Dr. Gabriel Cousens, MD, a psychiatrist and author, has commented that blue green algae increases mental alertness, memory and concentration in general, even in Alzheimer patients.

Blue green algae has an amino acid profile nearly identical to the amino acid profile of humans. Furthermore, according to Dr. Maurice Schiff, a professor of medicine at UCSD, blue green algae will chelate minerals proportionate to its amino acid pattern. Thus, this mineral pattern will combine together and arrange its mineral matrix as if it were identical to the building blocks of our very own flesh. The result? "Because everything is in its proper proportion, it is readily absorbed by the body. . ."[115] (see Figure 5.2).

These facts, i.e., that one species has a starch-based cell membrane, and that the mineral/nutrient matrix is so similar to humans, explain why blue green algae is up to 97% absorbed, assimilated and utilized when consumed.

Hormone Precursors. At least five bioavailable neurotransmitters (brain hormones) are found in significant quantities within blue green algae. Highly bioavailable quantities of neuropeptide building blocks (which are required to manufacture brain hormones) are also found in blue green algae. Furthermore, from studies utilizing blue green algae as a dietary adjunct in school children, it is now easy to recognize that nutritional status, physical health and mental ability are linked. Incredibly, consuming as little as one-half gram of algae daily dramatically improved Nicaraguan school children's overall nutritional profile, health picture and class attendance. Daily supplementation with blue green algae also dramatically raised test scores (see Figures 5.3 and 5.4).[b]

Thus, it appears from clinical observations and studies that blue green algae factors ensure an optimal balance of the hormonal precursors necessary to achieve efficient metabolism, both physical and mental.

Weight Loss. Blue green algae bolsters weight loss in two ways. First, its high concentration of protein and polysaccharides keeps blood sugar levels normal. This quenches the hunger apestat mechanism of the brain, regardless of intestinal fullness. Thus, the body is tricked into thinking that it is well sated. Second, blue green algae contains special protein precursors to neurotransmitters that repress the apestat control in

---

[b]Please see Appendix A for *The Genesis Effect: Super Brain Function and Restoration with Wild Blue Green Algae*.

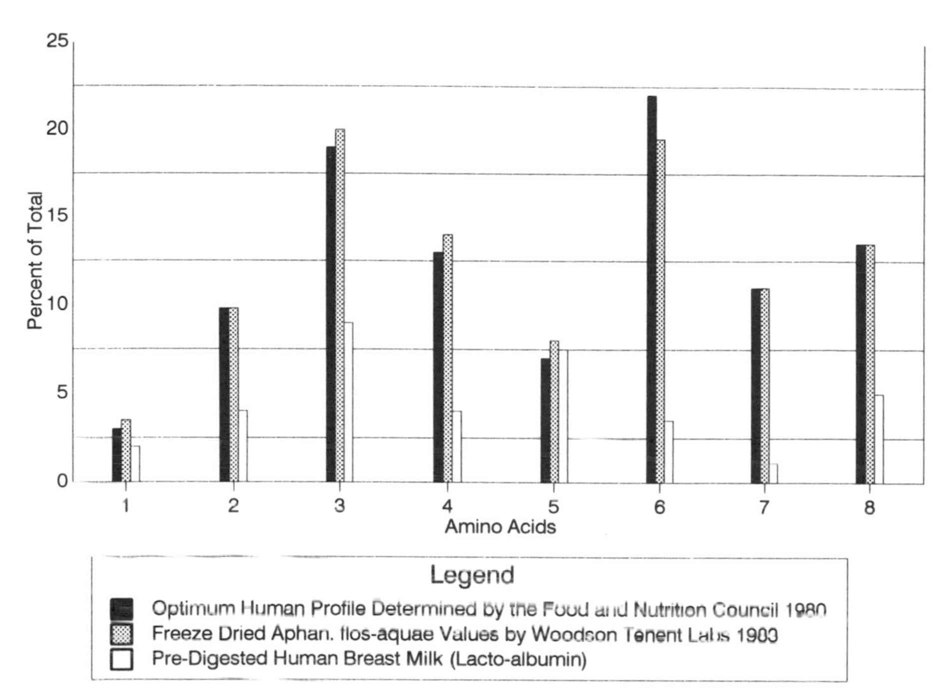

1. Histidine
2. Isoleucine
3. Leucine
4. Methionine plus Cystine
5. Phenylalanine plus Tyrosine
6. Threonine
7. Tryptophan
8. Valine

Figure 5.2 A Comparison of the Essential Amino Acids Found in Freeze-Dried *Aphanizomenon flos-aquae* and the Essential Amino Acids Found in Healthy Humans

Sources: Kollman, D. *Hope is a Molecule*, 1989 and Metagenics, "Pancreatic Digests (Di- and Tri-Peptides) of Proteins," professional handout.

the hypothalamus. According to Drs. William Goldwag and Marquetta Hungerford from the Center for Wholistic and Preventive Medicine in Stanton, California, this is a critical adjunct to successful weight loss. Dr. Richard Passwater stated that blue green algae specifically accomplishes this appetite suppression with its high content of the amino acid phenylalanine. In other words, this amino acid, because it is a precursor of norepinephrine, suppresses hunger by altering brain biochemistry in favor of the dieter.[116]

Recently, an important university study made a surprising discovery that dieting with a popular supermarket weight loss drink led to extensive muscle loss. In fact, it was found that over 70% of the weight lost in the

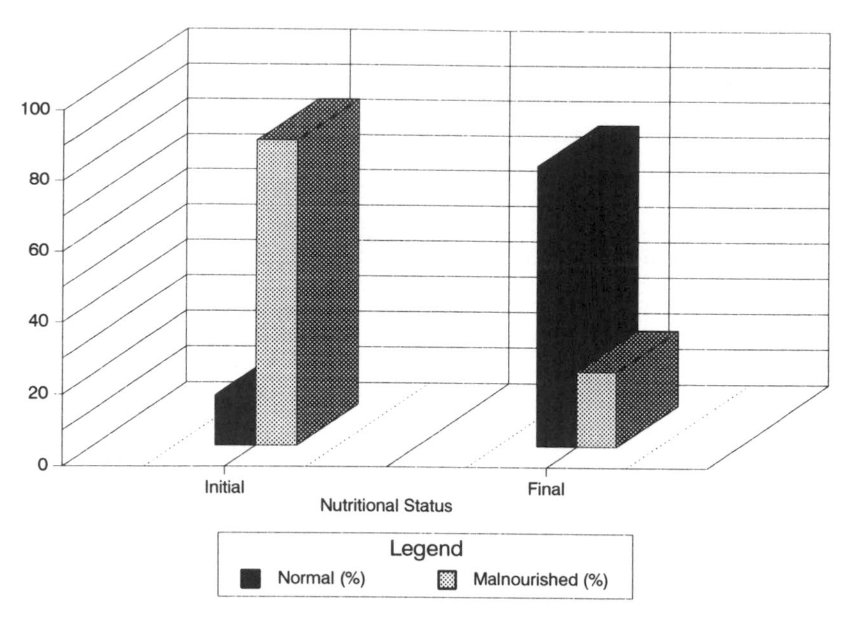

Figure 5.3 Change in Nutritional Status among Nicaraguan School Children

Source: Sevilla, I. and Aguirre, N. *The Nicaragua Report*, Special August Celebration Edition, Universidad Centroamericana, Nicaragua, May 1995.

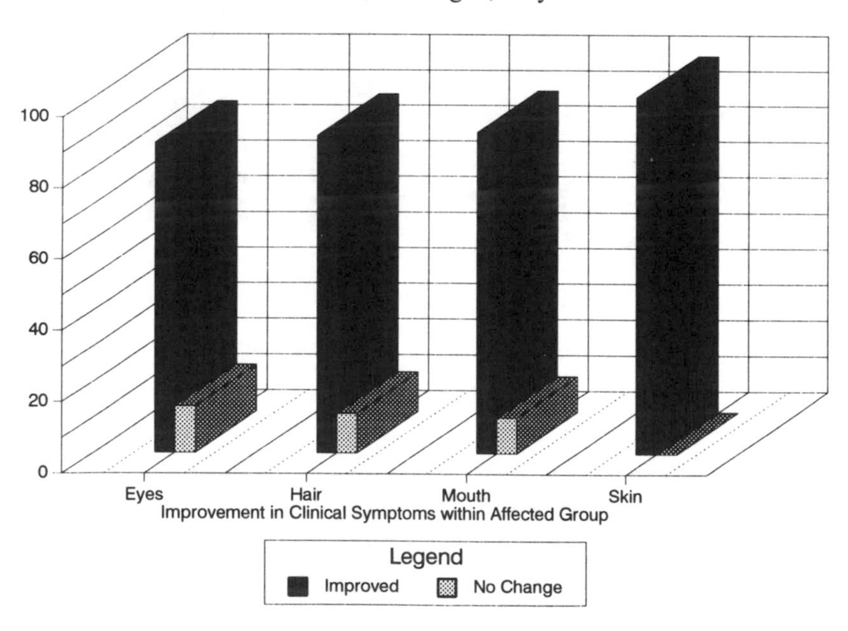

Figure 5.4 Change in Clinical Symptoms among Nicaraguan School Children

Source: Sevilla, I. and Aguirre, N. *The Nicaragua Report*, Special August Celebration Edition, Universidad Centroamericana, Nicaragua, May 1995.

supermarket diet group was vital muscle tissue. Such a dangerous type of weight loss must be prevented at all costs.

To increase muscle mass and lose weight at the same time is highly uncharacteristic for weight loss programs, yet very desirable for those trying to firm up and keep the weight off. Exercise during such dietary programs is essential. Most low calorie diet drinks like the popular supermarket diet drinks produce a lowered rate of metabolism. "This lowered metabolic rate often continues after dieting leading to quick weight gains," according to Dr. Doell. The blue green algae program, which contains large amounts of dietary RNA, is a powerful metabolic enhancer. Blue green algae contains unique amino acid sequences that with proper exercise, inspire muscle tissue production even during weight loss programs. Consuming blue green algae, with its rich and ideally-proportioned amino acid content, will help to spare organ and muscle mass during fat loss programs.[c]

Dr. Doell commented that, "muscle has a much higher metabolic energy demand than fat. Therefore, as the muscle mass increases, so does the body's thermogenic (heat-producing) responsiveness. With increased muscle mass and normal metabolic rate, the body burns more calories day to day reducing the common tendency to regain weight quickly after dieting."[117] Thus, when taken together, these unique nutrients offer great hope to people wrestling to maintain their ideal weight once the dieting program ends.[d]

Weight Gain. There is only one time in our life that our lean body mass (muscle, bone and organ tissue) will double in size and weight within six months. This only occurs during the first six months of life. The baby who is breast fed will attain this remarkable feat by suckling mother's breast milk with a mere 1.2% protein content! The unique protein structure in human breast milk is laden with branched chain amino acids, plus a uniquely human proportionate amino acid composition as previously elaborated.

Interestingly, blue green algae possesses a liberal amount of branched chain amino acids, among an amino acid pattern actually superior to human breast milk. These wondrous branched chain amino acids are: Leucine, Isoleucine and Valine. During exercise and growth, reserves of the branched chain amino acid Leucine can become critical, and therefore

---

[c]For further suggested reading see *Blue Green Algae and Super Sports Nutrition* (see Appendix A).

[d]For additional information see *The Genesis Effect: Perfect Weight Management* (see Appendix A).

rate limiting. Under greater metabolic demands, Leucine is typically gobbled up at an incredible 240% increase over normal periods of physical activity.[118] Therefore, it is crucial to have an abundance of dietary branched chain amino acids that can easily be supplied by the regular dietary addition of blue green algae.

In a fascinating study performed on Japanese school children, dramatic weight and height gains occurred after feeding 2 grams of chlorella daily for one-hundred twelve days.[119,120] Chlorella is actually classified as a "green algae" because it possesses cellulose within its cell walls. Cellulose is indigestible to humans, so chlorella must have its cell wall "cracked" in order to be readily absorbed. Therefore, care must be taken to prevent chlorella's internal exposed nutrients from spoiling. Shelf life can be shortened because of this situation.

Lastly, it is now well known that amino acid patterns consistent with those found in blue green algae, specifically the predigested protein constituency of human breast milk, are up to sixteen times more efficiently absorbed as compared to free-form crystalline (synthesized) amino acids.[121] Thus, this suggests that the best, most economical and most readily available ideal form of protein for the human body is blue green algae.

Active Enzymes. Blue green algae substances containing biologically-active enzymes that have not been inactivated by heating (or other) processes possess both anabolic (constructive) and catabolic (tearing down) capabilities. Eating freeze-dried blue green algae high in live enzymes should spare the body's inherent reserves of these precious life sustainers. "Live" food, in effect, digests itself via the action of these enzymes. Does the body have a finite source of digestive and other enzymes that can be squandered by an inappropriate dietary, stressful lifestyles or other factors? If our enzyme production capability is finite or is subject to external or internal interference, is it supportable and extendable by the use of the "live" enzymes contained in raw foods? In clinical practice, we see this answered in the affirmative time and time again.

Internally, Phase II detoxicating (see Cytosol Complexes p. 48) enzyme systems (many of which are natural constituents of blue green algae) include the following:[122]

1. Super Oxide Dismutase;
2. Catalase;
3. Glutathione peroxidase;
4. Methionine reductase;
5. And the critical conjugating peptide agents that encase toxins for removal:

a. Methionine;
b. Glycine;
c. Glutamic acid;
d. Taurine;
e. Serine;
f. L-Ornithine-L-Aspartate;
g. Histidine; and
h. D-glucuronic acid.

Super food concentrates that are rich in active enzymes are known to significantly raise blood enzyme levels within one hour after ingestion. Quantum Physicist and Biochemist Peter R. Rothschild, M.D., Ph.D., nominated for the Nobel Prize in Physics, has authored sixteen scientific books about enzymes. His latest research documents a newly-discovered method whereby enzymes taken orally significantly raise blood levels of those same enzymes within one hour. He demonstrated that oral enzymes may bond to emulsifying agents like lecithin, forming a liposome, which due to its low surface tension, can easily penetrate all intestinal mucosal layers and enter the internal lymphatic circulation.[123] It is known that RNA, a nucleic acid and the subject of our next section, forms readily into liposomes as well.[124]

Nuclear Complexes

Blue green algae contains approximately 4% nucleic acids and nuclear complexes, perhaps making it the richest commercially available source.[125] The nuclear complexes of the tissue cells include: (*a*) the nucleoproteins DNA and RNA for growth and repair; (*b*) nucleosides; (*c*) nucleotides; (*d*) the "helper" enzymes (primases) that help unravel or rewind and reorient the DNA/RNA templates for growth and repair; and (*e*) the energy-fueling molecule – adenosine tri-phosphate. At least twenty different unique proteins are involved in gene dynamics, activity about which very little is known.[126] This is partly because any form of laboratory analysis would necessarily disturb the complex properties of the regenerative apparatus. These natural constituents have never successfully been synthesized with the regenerative properties intact. However, it is possible to carefully extract them undisturbed with today's leading technologies (through freeze-drying methods) into whole food supplements. As previously pointed out, recent evidence suggests that these regenerative substances survive the digestive process and become available for use in tissue regeneration. Regenerative substances derived from blue green algae are vegetable sourced and affect tissue function and regeneration throughout the body in a generalized fashion.

Nuclear complexes are scientifically confirmed to induce: (*a*) unscheduled regeneration; (*b*) accelerated repair; (*c*) prevent, stop and even reverse many features of aging; (*d*) restore and greatly enhance immunity; (*e*) defeat cancer; (*f*) effectively neutralize radiation poisoning; (*g*) stimulate and return memory functions, even against advanced aging; (*h*) reverse organic brain disorders; (*h*) reverse genetic damage; and (*i*) serve as a par excellance adaptogen and stress modulator.[127,e] It is conceivable that our human genetic information may be augmented by blue green algae's ability to readily share its constituent nuclear complexes, prophages and plasmids.

Cytosol Complexes: Detoxification and Regeneration

Cytosol (cytoplasmic) complexes capable of invigorating and augmenting host tissue functional and structural regeneration represent another important area of study. These substances favorably enhance the support of debilitated tissues by supplying the necessities for functional and structural regeneration. Candice Pert, formerly with the National Institute of Health, has noted that some crucial cellular components are universal and interchangeable between species.[128] Detoxification and regeneration ultimately depend upon two crucial metabolic processes divided into two phases, called Phase I and Phase II Detoxication pathways. The P-450 system of Phase I Detoxication includes many forms of fully-activated vitamin B-complex and cyclic Adenosine Mono Phosphate (cAMP), that are entirely contained within blue green algae (see Water-Soluble Vitamins below). Phase II Detoxication is represented by critical enzyme systems (see *Active Enzymes*), at least partially present within blue green algae. By turning on, repairing and/or replacing the 5,000 or more enzymes found within each of our 27 trillion human cells, unexpected cleansing and deep scrubbing of our internal milieu ignites long lost youthful vigor and unscheduled regeneration.[f]

Chlorella has been heavily investigated for its potent detoxification of heavy metals, toxic chemicals, hepatotoxins and radioactive contaminants.[129,130,131,132,133]

---

[e]For a very complete and authoritative referenced article on RNA therapy as it relates to blue green algae, the reader may wish to order *Blue Green Algae and the Factors of Regeneration*, and *The Genesis Effect: Slowing, Stopping and Reversing the Aging Process*, from Genesis Communications (see Appendix A).

[f]For suggested reading toward a complete understanding of how cytosol complexes taken orally may induce regeneration in humans, the reader is directed to *Glandular Therapy*, Vol. I from Genesis Communications (see Appendix A).

Also included in the cytosol of blue green algae products are the following substances that are highly bio-available to promote functional regeneration:

Water-Soluble Vitamins. Blue green algae contains "matrix-bound" or "colloidal" water-soluble vitamins such as vitamin C and the coenzymated B-complex.[g] In the B-complex family, their most biologically-active forms for humans are represented by: Thiamine pyrophosphate, Cocarboxylase (vitamin B_1); FMN or FAD, Riboflavin-5'-phosphate (vitamin B_2); NAD[h] (vitamin B_3); Pyridoxal-5'-phosphate (vitamin B_6); 5'-Deoxyadenosylcobalamin (vitamin B_{12}); etc.[i] In contrast, the synthetic forms lack high-energy phosphorylation and are listed on vitamin packages simply as B_1, B_2, B_3, etc. Freeze-dried blue green algae appears to have high levels of the fully-activated energy rich forms of vitamins. On-going clinical studies illustrate the phenomenal health results shown by patients with vitamin deficiencies who consume blue green algae, verifying the potency of this tremendous food.[j]

Fat-Soluble Vitamins. Fat-soluble vitamins (A, E, F and K), very crucial fat complexes (Gamma linoleic acid, Glycolipids, Sulfolipids, Heptadecanoic and Myristic),[134] and the essential fatty acids in their most active forms are found within the cytosol components of blue green algae. The essential omega 3 and 6 fatty acids (i.e., gamma linoleic acid)[135] and their precursors (which form EPA) are also found in blue green algae.

The omega 3 and 6 oils govern hormonal balance and efficiency, as well as prevent and lessen many chronic degenerative diseases such as heart disease, diabetes, obesity and arthritis. In fact, blue green algae and human breast milk share a crucial feature – both contain significant amounts of gamma linoleic acid (GLA) (see Table 5.1).

---

[g]For a full discussion on matrix and colloidal vitamin forms, the reader is encouraged to read *Blue Green Algae and Vitaminology*. The superiority of natural nutrients over synthetics is also discussed in *The Genesis Effect: Mineralogy — Part I*. For more information see the Appendix A.

[h]NAD has recently been found to be important in reversing Type II Diabetes Mellitas.

[i]Please note that according to *Lancet*, Jan. 30, 1988; and *JAMA*, Dec 17, 1982, 80% of Spirulina's vitamin B_{12} content is bio-unavailable; whereas, other forms of blue green algae are not known to suffer from this defect.

[j]For more information, refer to *The Genesis Effect: Human Constitution Restoration* (see Appendix A).

Table 5.1 A Comparison of Vital Omega Oils between Breast Milk and Blue Green Algae

Activated Fatty Acid	Human Breast Milk	Blue Green Algae
C 18:3 (GLA)	0.34	0.25
C 20.3 (DGLA)	0.49	0.35
C 20.4 (AA)	0.77	0.47
C 20.5 (EPA)	0.43	0.30

[a]Fatty acid analysis of *Aphanizomenon flos aquae*. Alpha (Lot 1601DA) done for Dr. Barry Sears by Woodson Tenant Laboratories, Inc.
Source: Gibson and Kneebone, *Am. J. Clin. Nutrition* 32 353(1981)

Colloidal Minerals. Colloidal mineral complexes, i.e., RNA chelated minerals, matrix bound minerals and other highly-energized colloidal mineral configurations are found within blue green algae. In fact, blue green algae is approximately 22% natural mineral complex, much of this is bound to be in the colloidal matrix (i.e., enzyme bound) state. Such configurations cannot be synthesized. To date, over 40 macro and micro minerals have been identified in bluc green algae. The delivery rate of these minerals into the tissues is nearly 100%. Colloidal minerals and their actions as "determinants"[k] are required reading for any serious student of regeneration. The importance of food complexed minerals cannot be overstated. In 1936, top leading authorities [Dr. McCollum (Johns Hopkins), Dr. Mendell (Yale), Dr. Sherman (Columbia), Dr. Lipman (Rutgers), and Dr. H.G. Knight and Dr. Oswald Schreiner (USDA)], reported that, "Ninety-nine percent of the American people are deficient in these minerals and that a marked deficiency in any one of the more important minerals actually results in disease. Any upset of the balance, any considerable lack of one or another element, however microscopic the body requirement may be, and we sicken, suffer, shorten our lives."[136] Two volumes in *The Genesis Effect* Series are entirely dedicated to a comprehensive discussion on mineral colloidal states and the evolution of mineral dynamics.

Antioxidants. Blue green algae contains many of the known antioxidants such as vitamin C, bioflavonoids, beta carotene (one of the richest sources known), RNA, super oxide dismutase, catalase, gultathione, selenium, cysteine, etc. According to Dr. Charles Simone, author of

---

[k]A "determinant" is akin to the DNA molecule, in that it controls and dictates the processes of life.

Table 5.2 Beta Carotene Content of Foods

Food	Beta Carotene Content (in RE)*
Aph. flos-aquae (2g)	400.0
Carrots (1 whole, 70g)	360.0
Iceberg Lettuce (100g)	33.0
Tomato (1 fresh whole)	100.0
Apricots (5 small)	270.0
Eggs (1 hard cooked)	97.0 (retinol)

[a]1RE = 1 mc g retinol = 6 mcg beta carotene

Source: Data obtained from Kay, R.A., "Microalgae as Food and Supplement," *Crit. Rev. in Food Sci. and Nutr.*, vol. 30, 1991, p. 557. (by permission)

of *Cancer and Nutrition*, "If I had to recommend one nutrient above any other to include in a person's low fat, high fiber diet to reduce the risk of cancer, it would have to be beta carotene."[137] Today, an American would have to eat approximately three pounds of salad to equal the amount of beta carotene found in one teaspoon of blue green algae!

Associated Food Substances. Perhaps most importantly, a plethora of exciting associated food substances plus exciting unknown nutrient complexes yet remaining to be discovered are expected to be found within blue green algae. Some examples are several of the most biologically-active bioflavonoids (e.g., quercitin); mucopolysaccharides – indispensable for maintaining and repairing blood vessel walls and joints; cyclic Adenosine Mono Phosphate (cAMP) – a crucial energy-regulating modulator of many hormonal pathways; sulfolipids (Sulfoquinovosyl) – known to defeat viruses and viral related tumors;[138] unsaturated fatty acids – aid in reducing cholesterol levels; other cytochrome enzymes, co-enzymes and beneficial pigments (the newly discovered carotenoids – such as zeaxanthin and xanthophylls); and conjugating agents, i.e., choline and inositol.

One most exciting associated food substance is chlorophyll. Chlorophyll serves to order, balance, enhance and coordinate a plethora of critical metabolic functions. Thus, it serves as an electron-poising agent, protecting the very fabric of human metabolism. Chlorophyll serves as an effective antiseptic, oxygenator and detoxifying agent. The molecule chlorophyll can be readily bio-transformed into: (1) hemoglobin to rebuild blood; (2) vitamin B_{12} to insure optimal genetic expression; and (3) glucose tolerance factor to stabilize and even reverse hypoglycemia as well as adult onset (Type II) Diabetes Mellitus.[139]

Chlorophyll has been scientifically documented to successfully treat ulcers, infections, anemia, hypoxia, stagnated wound repair, toxemia, severe burns and more.[l]

Another incredible associated food substance is chlorella growth factor. Similar to epithelial growth factor, chlorella growth factor actually induces unscheduled repair, making it a potent regenerative agent. Mysteriously, chlorella growth factor is showing promise at inhibiting cancer growth. The proposed rationale for this exciting effect may lie in its high concentration of nucleic acids and their building blocks: adenosine and cytidine.[140]

The glycoproteins found in blue green algae have been cited for powerful anti-cancer properties as well.[141] *Aphanizomenon flos-aquae* is another species of blue green algae rich in glycoproteins (a constituent of its cell membrane).

Glycolipids, also found in blue green algae, have gained world-wide attention as immune modulators and stimulants. A National Cancer Institute study demonstrated blue green algae's glycolipid content to be remarkably effective in suppressing the AIDS virus in the test tube.[142]

Nutritional Analysis and Comparison of Blue Green Algae

See Table 5.3 for a complete listing of the nutritional compositions of various forms of blue green algae.

Undesirable Constituents of Blue Green Algae

Because of the startling rise in popularity in the use of blue green algae, many critics unfamiliar with this super food have issued hypervigilant concerns. This section places such concerns into their proper perspective. For a complete listing and elaboration of techniques employed by manufacturers and suppliers of blue green algae, the reader is directed to *Blue Green Algae: Production and Safeguards*.[m]

---

[l]For a complete and authoritative referenced article on chlorophyll therapy as it relates to blue-green algae, the reader is again referred to *Blue Green Algae and the Factors of Regeneration* (see Appendix A).

[m]See Appendix A for ordering information.

Table 5.3 Nutritional Value of Commercially-Preproduced Food Grade Microalgae

	Aph. Flos-Aquae[a]	*Chlorella*[a]	*Spirulina*[a]
Ash	7%	3%	7%
Carbohydrate	27%	23%	18%
Moisture	6%	5%	5%
Nucleic Acid	4%	3%	4.5%
Protein	60-69%	60%	65%
Total Lipid	3%	9%	5%
Minerals			
Calcium	140.0 mg	22.0 mg	100.0 mg
Chlorine	46.0 mg	n/a	44.0 mg
Chromium	40.0 mg	n/a	28.0 μg
Copper	60.0 μg	10.0 μg	120.0 μg
Iron	6.4 mg	13.0 mg	15.0 mg
Magnesium	16.0 mg	32.0 mg	40.0 mg
Manganese	0.3 mg	n/a	0.5 mg
Phosphorus	51.0 mg	90.0 mg	90.0 mg
Potassium	100.0 mg	90.0 mg	120.0 mg
Sodium	38.0 mg	n/a	60.0 mg
Zinc	0.3 mg	7.0 mg	0.3 mg
Vitamins			
Ascorbic Acid (vit. C)	5.0 mg	1.0 mg	0.5 mg
Biotin	3.6 μg	19.0 μg	0.5 μg
Carotene	2,000.0 RE	550.0 RE	2,300.0 RE
Choline	2.6 mg	n/a	n/a
Cobalamin (vit. B_{12})	8.0 μg	1.3 μg	3.2 μg
Folic Acid	1.0 μg	2.7 μg	1.0 μg
Iositol	n/a	13.2 mg	6.4 mg
Niacin (vit. B_3)	0.65 mg	2.38 mg	1.46 mg
Pantothenic Acid (vit. B_5)	130.0 μg	130.0 μg	10.0 μg
Pyridoxine (vit. B_6)	67.0 μg	140.0 μg	80.0 μg
Thiamin (vit. B_1)	0.03 mg	0.17 mg	0.31 mg
Vitamin E	1.2 IU	0.1 IU	1.0 IU
Associated Food Substance			
Chlorophyll	300.0 mg	200.0 μg	115.0 mg

[a]Figures are given per 10g dry weight.

Source: Data obtained from Kay, R.A., "Microalgae as Food and Supplement," *Crit. Rev. in Food Sci. and Nutr.*, vol. 30, 1991, p. 557.

Rancid Fats

Blue green algae nutritional products must be free of rancid fats that pose a threat to normal metabolism. Rancidity is caused by contact between blue green algae's intrinsic fat content and oxygen during preparation. Chlorophyll is fat soluble. Rancid by-products of chlorophyll may include chlorophyllides, pheophorbides and pyropheophorbides. Never in any reported testing of blue green algae products destined for the consumer have these toxic by-products exceeded 0.2%, which falls well within the most stringent standard. Spray-drying techniques are particularly susceptible to this form of rancidity. This is due to the fact that air contains oxygen, and as the algae is ejected at extremely high pressure into air, the algae becomes blasted through and through with oxygen. Hence, some rancidity must necessarily result.

Environmental Toxins

The end product must be free of any detectable amounts of environmental poisons, such as lead, methyl mercury, pesticides and water or air-borne pollutants that are easily absorbed by blue green algae. This demands use of an effective filter process and source material from bodies of water that prohibit the use of or are totally devoid of environmental toxins. One well known bio-mass rich in blue green algae is Upper Klamath Lake, Oregon. Upper Klamath Lake is known to be totally unpolluted. No pollutants have ever been found in blue green algae harvested from this lake. Many salt water sources used for blue green algae cultivation have also never demonstrated pollutants of any type.

Pathogens

No pathogenic disease-associated bacteria of any type is to be present in the final end-product. Freeze drying accomplishes cold sterilization, as does emersion in alcohol, and hence cannot contaminate or cause infection of any type.

Labeling Claims

The end product must meet or exceed the labeling claims of the manufacturer.

Additional and Proposed Guidelines

The following guidelines are recommended for all suppliers:

1. *Independent Laboratory Evaluation* on a random basis. The Linus Pauling Institute is one independent company that conducts these tests. At least one major distributor of blue green algae currently employs random evaluations.
2. *The Product's Rawness* for blue green algae should be confirmed by a minimum of three criteria:
   - The total protein coagulability factor.
   - Total vitamin C content (intrinsic sources).
   - At least two or more specific enzyme tests, and one appropriate generic enzyme test.
   - Total chlorophyll-magnesium complex content.
3. *Nucleoprotein Content* as well as nucleic acid, nucleoside and nucleotide content (quantity).
4. *Assays of Adenosine Tri-Phosphate Content.*
5. *Assays of Toxic Metabolites* such as the disintegrating by products emanating from rancid chlorophyll.

Dosages and Expected Results

There are basically three levels of experiences people will have when consuming high-quality blue green algae. Approximately one-third of first-time users will get an immediate lift and/or exhilaration. This usually occurs within three days of starting the product.

Another one-third of first-time users will experience mild to major symptoms of detoxification with high-quality blue green algae. Depending on whether the person is congested with simple water-soluble toxins or inundated with fat-soluble toxins, highly rewarding experiences of well being may not occur for three weeks, or even as long as three months.

And, in the final one-third of first-time blue green algae consumers, no immediate results, one way or another, will be felt. It is important that this group experiment with dosage. Usually by doubling intake every five to seven days, at some point these "no responders" finally start responding. The response will be either a very noticeable sense of well being or detoxification.

Thus, the regenerative actions of blue green algae occur at the three day mark, three- to four-week mark, and/or three- to four-month mark. The best "benchmark" to assessing the benefits of high-quality blue green algae lies in its ability to evoke detoxification reactions. The key to understanding this phenomenon of blue green algae is to understand that continued consumption of blue green algae almost always eventually clears all detoxification signs and symptoms. If blue green algae were actually fooling us, in other words, actually causing toxic symptoms, then its continued use would certainly not result in a lessening of symptoms, nor lead to positive experiences of well being. Yet, this is exactly what consumers of high-quality blue green algae often observe and experience.

Careful selection of dosages is essential to achieve optimal results. Some people feel wonderful on as little as 1 gram of blue green algae daily. Others may require up to 10 grams daily to "power up." Dosages should always start small, gradually building up at four- to seven-day increments.

Dosages

When starting a blue green algae program, it is important to consume one-half gallon daily of purified water. Start with one tablet (250 mg) of high-quality blue green algae three times daily with meals. Do this for six days. On the seventh day, add in one more tablet of blue green algae with each meal. Do this for one week more. These tablets begin the process of body and nerve invigoration. On day fifteen, double the amount. For example, take four blue green algae (1,000 mg) tablets with each meal.

Detoxification Symptoms

Being evaluated by way of a health questionnaire may predict if you are likely to experience discharge symptoms. Early signs of detoxification suggesting that you are not drinking enough water, or if there is an unusual backlog of internal congestion, include: (*a*) mouth soreness; (*b*) stiffness; (*c*) mental fog; and/or (*d*) temporary skin eliminations.

Eleven percent of patients will go on to experience more intense symptoms during the initial four weeks of blue green algae supplementation, but these rapidly and uneventfully disappear. These fleeting, but discomforting symptoms are signs your body is coming alive and discharging congestion. It is a good thing, but inform your wholistic health care practitioner immediately so that he/she can help you to pass through this brief period as comfortably as possible. These include: (*a*) nausea (most common); (*b*) achy, flu-like symptoms (second most common); (*c*) headaches (third most common); and (*d*) diarrhea (least common).

These detoxification reactions are temporary and are experienced by a few people. If these occur simply: (*a*) take a one-day break from your blue green algae program; (*b*) be sure you are drinking enough pure water; and then (*c*) gradually wean yourself back onto the protocol from the beginning again. In this way, you minimize chances of encountering future detoxification reactions. Whenever symptoms of detoxification discomfort resurface after resting all blue green algae intake for one full day or more, the patient must stop all intake of blue green algae and substitute specifically-designed support products (i.e., acidophilus and super sprouts). These support products are given for one whole week before re-administering the pure algae products. The support products are then continued until the bottles run out. Again attempt to re-wean back onto the above schedule from the beginning, unless symptoms return, until you reach full dosages. If detoxification reactions resurface at any point along the way, lower intake to the amount that caused no noticeable symptoms and stay at that level for two more weeks, then try increasing back to maintenance dosages; i.e., day fifteen.

Follow these instructions for support product intake: (*a*) take two enzymes with each meal; (*b*) take three acidophilus immediately upon rising; (*c*) take two sprout tablets with the acidophilus in the morning plus two more just prior to bedtime.

To play conservative, many health advisors recommending blue green algae supplementation prefer to have clients start out on support products first to avoid or lessen the chances of possible detoxification reactions. By using a sophisticated health questionnaire, highly accurate and individualized programs can be created for most people. These programs are simple and the reader is referred to Appendix A under the, *Genesis Adaptation Survey — (GAS)* form discussed in volume 8.

Conclusions

Blue green algae has adapted to growing in pools of water located in the desert. Occasionally, when these waters evaporate, blue green algae turns a frosted white. Under these conditions, blue green algae is transformed into complex sugars which result from sunlight altering its 65% protein content. There has been speculation that the hidden manna of the wandering Israelites (see Exodus 16) that appeared miraculously on desert rocks following a devastating dry spell may have been the result of blue green algae transformation. Blue green algae was discovered in Lake Chad in 1964 by a Belgium botanist named Jean Leonard.[143]

> When the Spanish invaded Mexico nearly 500 years ago, they discovered Aztec Indians surviving on a mysterious blue green algae growing on Lake Texcoco. Named "tecuitlatl," it was used by the Aztecs as a daily food source and trade item. . . . some archaeologists speculate spirulina enabled the culture to thrive and flourish. Aztec legends reportedly tell of messengers who took spirulina on their marathon runs.
>
> . . . Guatemala's Mayan Indians may have cultivated it in elaborate waterways uncovered by modern archaeologists. . . . Many archaeologists believe the waterways were built for algae maintenance pools. . . .
>
> If algae farms existed in the Mayan culture, it would explain how the Mayan culture sustained a population of two million people during the height of the Classic Period (900 A.D.), despite depleted agricultural resources.
>
> The Kanembu tribe in Africa's Sahara Desert collects spirulina from Lake Chad and dries it into hardened cakes called "dihe," which are sold in native markets. The Africans crumble the dihe and mix it with a spicy tomato sauce that's poured over grains and meat.
>
> Because algae is highly nutritious and easily cultivated in small spaces, the U.S. National Aeronautics and Space Administration (NASA) is investigating it as a food for humans aboard permanent space stations. . . .
>
> The NASA studies involve not only spirulina, but also chlorella and. . . (*Aphanizomenon flos-aquae*).[144]

In conclusion, blue green algae is the *first food*, *earth's major food*, and perhaps the *most abundant whole food rejuvenator* still available to modern civilization. Those who partake of the incredible benefits of blue green algae as a dietary addition will be enormously indebted to this miracle of life.

Genesis Quicknotes

1. The bankruptcy of America's soil is responsible for most of our ill health.
2. Nearly all Americans are ill.
3. Algae has been consumed as part of the human diet worldwide throughout the ages.
4 Blue green algae may be the most potent nutritive and regenerative food known for human beings.
5. Algae's influence on planet earth is unexcelled. Therefore, in order to overcome America's health crises, what better ally could we have than blue green algae?

Appendix A

The Genesis Paradigm

A few of the most remarkable healers of our age, Dr. Max Gerson, Dr. William Kelly, Dr. Nicolas Gonzalez, Dr. William Powell Cottrille, and Dr. George Goodheart, have delineated reproducible methods to induce Quantum healing in their patients. At first, the method simply encompassed four issues:

1. Detoxification
2. Optimal oxygenation
3. Eating abundant raw food factors
4. Correcting structural stress, tension and misalignments

Then, starting in 1945, this list necessarily increased due to the post WWII era and the onset of the nuclear age. At this point in time, unprecedented toxins made their debut in our world, and were comprised of previously unknown forms of hideous poisons, such as:

1. radioactive wastes,
2. chemicals,
3. metals, ... and soon, more and
4. more formidable infectious agents,

all of which often become most unwanted and fixated house guests.

Therefore, the method to consistently reproduce successful regenerative experiences evolved to include techniques capable of handling the newer anti-health/anti-planet arrivals. The following discussion will elaborate on this novel approach, and the paradigm that it begets will be found throughout *The Genesis Effect* volumes referenced respectively.

Detoxification – "The more powerful the regenerative properties are within a super food, the more likely, persistent and intense the detoxification activation will be."

Detoxification is the first and most important pre-requisite to regeneration. The more knowledge that the student acquires on the subject on detoxification, the more successful the regenerative achievement will be. Detoxification must be managed by both reality and experiential insight.

The less one knows on the subject of detoxification, the more the confusion and resistance will manifest. Be aware that a diamond is nothing more than a lump of coal that stuck with it. But also be aware that follow through and persistence will expedite your first and every step on the journey of a thousand miles. In this way, you are ultimately bound to be successful.

See volumes 1, 2, 4, 6, 8, 9, 15, 17

O_2 – Oxygen, the spark of life. You can live for forty days or more without food; seven days or more without water, but only seven minutes without oxygen. Thus, oxygen is truly the most important nutrient of all. Some of the greatest Noble Laureates' sole inspiration emanated from this, creation's most precious breath. To super saturate with oxygen is to maximize regenerative functions, but protection for this finest of flames must simultaneously be in place. Thus, the subject of antioxidants (nutrients which direct the best and least wasteful use of oxygen) is required reading for the serious student of regeneration.

See volumes 1, 2, 3, 6, 9, 11

Embryonic Food Factors/*Genesis Effectors* – Are only found in super foods. Super foods are only super foods when they contain powerful factors of regeneration.

See volumes 1, 2, 3, 7, 9, 1, 13

Subluxation-Dysponesis: CNS/ANS/pH – A paradigm shift in our understanding of health is necessary today more than ever. The linear model of medicine is now made obsolete by the non-linear, multifactorial sicknesses, disease syndromes and chronic illnesses now running unchecked and in epidemic proportions in the civilized world. The subluxation complex is interference with the body's optimal health expression. As Hippocrates said, the physicians's only role is to unleash this inherent self-healing power. The physician should not be consumed via external means, to destroy the disease process itself, lest it shall sooner than later come back to haunt.

See volumes 14, 15

May you all share in the many splendid benefits of *The Genesis Effect*. Aim high so that your results are nothing short of real regeneration: longevity, constitutional renewal and global renewal.

Books Available from Genesis Communications

1. ***The Genesis Effect: Spearheading Regeneration with Wild Blue Green Algae.***

This first volume offers the reader excellent insight into the world of super foods and the regeneration paradigm. Blue green algae figures prominently in this discussion, complete with an explanation of why, where and how America is losing its health; and how people, like you, are regaining their past heritage of optimal wellness. Ninety-six pages, fully referenced, perfect bound, price = $7.95 ea. Bulk discounts available.

2. ***Blue Green Algae and the Factors of Regeneration***

A great short synopsis of the reality of regeneration with blue green algae. 8 pages, fully referenced, price = $2.00, bulk discounts available.

3. ***The Genesis Effect: Super Brain Function and Restoration with Wild Blue Green Algae***

This volume is all about restoring optimal brain function, even in the face of such devastating illnesses as senile dementia, Alzheimer's and Attention Deficit Hyperactivity Disorder (ADHD).

4. ***Blue Green Algae and Vitaminology***

The myth that "A vitamin is a vitamin, is a vitamin" is shattered in this equally shattering exposé on the synthetic vitamin industry.

5. ***The Genesis Effect: Addiction Elimination***

The hidden hunger phenomenon is the essence behind all addictive processes, and no amount of money spent on mental health will ever address the cause of addictions until super foods are used to quench this form of starvation.

6. ***The Genesis Effect: Perfect Weight Management***

Stop losing weight with diets predisposing you toward gaining it all back again. This process is enormously harmful and should be outlawed! Fast fix means fast highway to body/mind bankruptcy.

Weight loss, weight gain and sports are all more related than you think. Find out how everyone can benefit from super nutrition and organ and muscle regeneration techniques.

7. *Blue Green Algae and Super Sports Nutrition*

For athletes desiring super performance and fitness, this is a must book.

8. *Genesis Adaptation Survey (GAS): Handbook to Detoxification, Patient Management and Clinical Findings*

Managing health recovery by reality is no simple task, or is it? This volume dispels the myth that monitoring regeneration is too complicated or extravagant. Contained herein are simple wellness surveys that chart progress and predict hurdles so that a unique game plan can be easily created for each individual.

9. *The Genesis Effect: Slowing, Stopping and Reversing the Aging Process*

Longevity in total health is not just a dream, it is a reality. This volume reveals the well-documented facts about how we may all achieve a greater and healthier lifespan.

10. *Blue Green Algae: Production and Safeguards.*

Rumors abound among algaphobes regarding certain forms of edible algae. This volume smashes through the veils of innuendo and algae bashing occasionally encountered in the media and with friends.

11. *The Genesis Effect: Immunity and Genesis Cybernetics*

Finally, a total solution to the immunity issues devastating modern civilization.

12. *The Genesis Effect: Mineralogy — Part I*

From dust to dust do we enter and part from this world. Minerals are not what you think. There is indeed a consciousness guiding their every state and form. They are everything. Minerals control all aspects of nutritional programs. They also control the planets' health as well. Thus, our link to the environment is the most important lesson to learn in our total body/mind health recovery.

13. *The Genesis Effect: Mineralogy & Enzymology — Part II*

Enzymes work ten-thousand times faster than all other super nutrients. Certain minerals inhibit enzymes, others are required for enzyme activation. Minerals control the pH and pH controls all enzymes. That should leave a lasting impression.

14. *The Genesis Effect: Human Constitution Restoration*

Putting all this together is now made possible by resurrecting the forgotten schools of constitutional analysis. This volume presents a fascinating discussion on defining our true birth right of optimal wellness and longevity.

15. *Beautiful Skin with Wild Blue Green Algae*

It has been known for decades that the major constituent of Blue Green Algae, RNA, has been shown to significantly reduce wrinkles, skin blemishes and acne if taken in adequate amounts for adequate periods of time. This short fascinating synopsis covers all the bases when it comes to attaining healthful, beautiful and more glowing complexion.

16. *The Genesis Effect: Spiritual Renewal*

Here we look at communing with our own spiritual growth and development before we project outwardly. "Physician heal thyself first, and then you will be able to heal the sick." This biblical reference is no where more apropros than for this volume's focus. We can all find our way spiritually while we selflessly serve the needs of another's spiritual growth. In this fashion, there is no element of conversion, just simple inspiration by setting the right type of examples.

17. *The Genesis Project: The Global Vision for Planetary Renewal*

A total local, regional, national and global actualization of renewal is suggested in this volume based on your experiences with the previous volumes. Subjects broached will included the founding of: (1) health insurance that covers alternative methods to clients demonstrating responsibility for their own health; (2) health malpractice insurance for natural health care practitioners; (3) a legal trust fund available to eligible health practitioners under attack; (4) special energy-efficient housing for apartments for a welfare rehabilitated workforce (In the past, the government has paid a premium for employment agencies that find jobs for the welfare class); (5) organic gardens placed behind such units for children to be gainfully employed; (6) fast food restaurants that would utilize organically-grown foods in their menu selections, along with BGA milkshakes and dishes; (7) composting garbage dumps and recycling centers that capitalize on recycling America; (8) schools that super nourish the students with food for mind and body; (9) the Genesis Health Center – a complete natural health care facility specializing in regenerative protocols and serving in the dual capacity as a teaching and research center. The center will be located in Hawaii, with satellite locations throughout the United States being a primary focus of growth.

Appendix B
The Power of Networking

For the second time in 18 years on Saturday, October 8th, 1994, the U.S. Congress passed and President Clinton signed into law, a most powerful pro-health law concerning nutritional supplements. The first time was when Senator William Proxmire sponsored a bill passed into law in 1976, which protected the nutritional supplement industry's product availability as we know it today. We have much to thank him for; however, some key areas were left unaddressed. The new bill, which attempted to address those problematic areas, was called the Hatch-Richardson-Harding Bill (S.784). This new law allows any manufacturer or representative of a dietary supplement to make truthful and non-misleading statements as to how dietary ingredients affect human body functions and structure. There must be a similarly worded FDA disclaimer as written in the front section of this volume under "Special Reader Advisory" whenever a manufacturer elects to disseminate proper educational materials on nutritional ingredients relevant to nutritional products being sold by that manufacturer.

Furthermore, third-party literature can be used that presents a balanced view, such as peer-reviewed scientific studies, that educate the reader about the truthful merits of nutritional ingredients. The only restriction on this educational discussion is that it completely avoids promoting a specific company name or a specific product for promotional purposes.

Additionally, the burden of proof that a specific dietary supplement is harmful and possesses a significant **"or"** unreasonable risk to the consumer would lie squarely on the shoulders of the FDA, and *not* the manufacturer. Now, the FDA must first go to court to *prove* their case before taking any further action against any manufacturer. This suggests that the law is intended to portray nutrients as *innocent* until proven guilty (i.e., truly harmful) by the FDA. Previously, the antiquated system used by concerned federal agencies and special medical interests (long overdue for an extensive overhaul) insisted that all nutritional supplements were suspect until many years of thoroughly redundant testing proved otherwise, sometimes for as long as 50 years or more (kindly refer to Volumes 12 and 13 of *The Genesis Effect* series). Thus, finally the federal government has awarded citizens the rights to use natural nutrients, the same inalienable rights guaranteed under the U.S. Constitution to every American citizen. The voice of the people has been heard, louder and clearer than ever, thanks to the unrelenting and talented Senators sponsoring and insuring the passage of bill S.784.

However, because the word "or" was used instead of "and," between the words "significant or unreasonable," some are concerned that the word "significant" as a stand-alone criterion is not sufficiently defined under this new law. For example, if various allergic reactions, or for that matter, *significant detoxification symptoms*, were to arise after consuming a dietary supplement, would the FDA invoke a court review on this aspect of the law, declaring that significant risk to the consumer required that the FDA take the product off the market? This is uncharted territory. **Therefore, this book was in part written to educate the reader about the benefits of normal detoxification reactions, and what a logical course of action would be if and when detoxification was experienced.** More importantly, because of the previous law, more than a few manufacturers have maintained exhaustive independent analysis of their product line to insure the highest possible quality assurance standards. And finally, in all circumstances where a question arises as to the possible adverse effects of a nutritional supplement (such as a possible allergic reaction), a properly-trained health care practitioner should be immediately consulted.

Under such protocols, i.e., that:

(a) the potential consumer is adequately educated and familiarized with natural healing's occasional detoxification (Herxheimer) manifestations;

plus

(b) the manufacturer employs state-of-the-art safeguards and exhaustive quality assurance analysis of all products;

and

(c) the consumer judiciously consults with an adequately-qualified healthcare practitioner whenever and wherever necessary;

the need for the FDA to prematurely invoke court review over a dietary supplement issue is greatly minimized.

Certainly, this series of volumes relating to the scientifically-documented benefits of blue green algae, was written with all of these stipulations firmly in mind.

Super foods such as food grade blue green algae are foods, so their regulation falls under the auspices of the United States Department of Agriculture (USDA) and not the FDA. Interestingly, this aforementioned S.784 bill also created an exciting new Office of Dietary Supplements within the National Institutes of Health (NIH). This further relaxes the role of the FDA as absolute "overseer" to the health food industry. The FDA would still step in to "regulate" a food or nutrient if it was being represented by a manufacturer or agent of a nutritional company as a substance that treats a specific medical disease, or if a significant or unreasonable risk was proven in court by the FDA to exist in the consumption of a nutritional supplement.[a]

In conclusion, what the author does indeed wish to express is the simple concept that only the body is truly capable of curing "naturally" any disease, and that the super nourishment found within blue green algae offers the "par excellence" support the body needs to regenerate tissues and recover optimal wellness wherever and whenever possible.

---

[a]Note: Many readers make the mistake that the FDA "approves" various items. The FDA is not empowered to "approve" anything, only to *regulate* within applicable law. Therefore, the author insists that no person should use the information contained herein to prescribe a treatment for a medical condition, or to make the false claim that the FDA has "approved" anything relating to a health product.

References

1. Smith, R. *Where Is the Wisdom: The Poverty of Medical Evidence*. The British Medical Journal, vol. 303, 1991, pp. 798-99.
2. *The New Webster's Dictionary*, revised ed., Modern Publishing, New York, NY, 1987, p. 258.
3. *The Reader's Digest Great Encyclopedic Dictionary*, The Reader's Digest Association, Inc., Pleasantville, NY, 1966, p. 1132.
4. Albrecht, W.A., "Our Teeth & Our Soils," *Annals of Dentistry*, vol. 6, no. 4, December 1947, p. 199-213.
5. "Diseases as Deficiencies via the Soil," *The Iowa State College Veterinarian*, vol. 12, no. 3, Iowa State College.
6. Allison, I., "Are We Starving At Our Tables?" *Steel Horizons*, vol. 12, no. 3.
7. Albrecht, W.A., "Our Teeth and Our Soils," p. 209.
8. Ibid., p. 199-213.
9. Price, W.A., *Nutrition and Physical Degeneration*, Heritage ed., The Price-Pottenger Nutrition Foundation, Inc., La Mesa, CA, 1982, p. 21.
10. Sheets, O., "The Relation of Soil Fertility to Human Nutrition," Miss. Agr. Expt. Sta. Bull. 437, 20 p., 1946.
11. Ibid.
12. Ebeling, W.,"How Fertilizers Affect the Nutrient Balance in Plant Crops," *J. Appl. Nutr.*, vol. 33, no. 2, Fall 1981, p. 138-39.
13. Duncan, C.W., "Effects of Fertilizer Practices on Plant Composition. Field Results." In *Nutrition of Plants, Animals, and Man*, Mich. St. Univ. Col. Agr., Centennial Symposium, East Lansing, MI, 1955a, p. 20 26.
14. Sheldon, V.I., et al., "Diversity of Amino Acids in Legumes According to the Soil Fertility," *Science*, vol. 108, 1948, p. 426-28.
15. Ebeling, W., "How Fertilizers Affect the Nutrient Balance in Plant Crops," p. 143-45.
16. Hopkins, H.T., et al., "Soil Factors and Food Consumption," *Am. J. Clin. Nutr.*, vol. 18, 1966, p. 390-95.
17. Howard, A., "Natural v. Artificial Nitrates," *Organic Gardening*, August, 1945, p. 7.
18. Stoff, J.A. and Pellegrino, C.R., *Chronic Fatigue Syndrome: The Hidden Epidemic*, Random House, Inc., New York, NY, 1988, p. 113-14.
19. Murray, R.P., "Obesity: A Disease of Starvation," *The Clinical Nutritionist Newsletter*, vol. 3, no. 9, Sept., 1983, p. 328.
20. Price, W.A., *Nutrition and Physical Degeneration: A Comparison of Primitive and Modern Diets and Their Effects*, Heritage ed., The Price-Pottenger Nutrition Foundation, La Mesa, CA, 1970, p. 19-20.
21. An NBC report, "The Almond Brothers," given on the NBC Evening News, December 7th, 1989.

22. Ebeling, W., "The Relation of Soil Quality to the Nutritional Value of Plant Crops," *J. Appl. Nutr.*, vol. 33, no. 1, Spring 1981, p. 24-34.
23. Carrel, A., "Tissue Culture and Cell Physiology." *Physiol. Rev.*, vol. 4, 1924, p. 1-17.
24. Ibid.
25. Lee, R. and Hanson, W., *Protomorphology: the Principles of Cell Auto-R*, Lee Foundation For Nutritional Research, Milwaukee, WI, 1947, p. 56.
26. Carrel, A., "Artificial Activation of the Growth in Vitro of Connective Tissue," *J. Exp. Med.*, vol. 17, 1913, p. 14-19.
27. Carrel, A., "Tissue Culture and Cell Physiology."
28. Carrel. A., *JAMA*, Jan. 26th, 1924, p. 256-57.
29. Carrel, A., and Ebeling, A.H., "Heat and Growth-Inhibiting Action of Serum," *J. Exp. Med.*, vol. 35, 1922, p. 647-56.
30. Carrel, A., *JAMA*, Jan. 26th, 1924, p. 256-57.
31. Starzl, T.E., et al., "Growth Stimulating Factor in Regenerating Canine Liver," *Lancet*, vol. 127, 1979.
32. Makowka, L., et al., "The Effect of Liver Cytosol on Hepatic Regeneration and Tumor Growth," *Cancer*, vol. 51, 1983, p. 2181-90.
33. Das, M., "Epidermal Growth Factor Receptor and Mechanisms for Animal Cell Division," *Current Topics in Membranes & Transport*, vol. 18, 1983, p. 381.
34. Fisher, "Nature of the Growth Promoting Substances in the Embryonic Tissue Juice. A Review of the Author's Investigations," *Acta Physiol. Scand.*, vol. 3, 1941, p. 54-70. (Also see *Nature*, July 21st, 1962.)
35. Newman, E.A., et al., "Effect of Nucleic Acid Supplements in the Diet on Rate of Regeneration of Liver in Rats," *Am. J. of Physiol.*, vol. 164, 1951, p. 251.
36. Dimitriadis, G.J., "Introduction of Ribonucleic Acids into Cells by Means of Liposomes," *Nucleic Acid Research*, vol. 5, 1978, p. 1381.
37. Babich, F.R., et al., "Transfer of a Response to Naive Rats by Injection of Ribonucleic Acid Extracted from Trained Rats," *Science*, vol. 149, 1965, p. 656-57.
38. Wacker, A., "Beeinflussung der Protheinsynthese Durch in Vivo-Verabreichung von Organspezifischen RNS-Praparaten," *Personliche Mitteilung Vom*, 7.7.71 (unveroffentlichte Versuche).
39. Axxman, G., "Untersuchungen zur Organotropen wirkung von Zellularen Extrakten auf die Proteinsynthese in Vivo," *Diplomarbeit Vom Marz*, 1973, Universitat Frankfurt, Institut fur Therapeutische Biochemie.
40. Bethge, J.F.L., et al., "Versuche zur Verkurzung der Frakturheilungszeit. III. Ribonukleinsauren," *Langenbecks Arch. Chir.*, vol. 333, 1973, p. 153-64, (Chirurg. Univ. -Klinik und -Poliklinik, Hamburg-Eppendorf).
41. Greife, H., et al., *Tierernaehr. Futtermittelkd.* vol. 40, no. 5, 1978, 248-56. (inst. Tierphysiol. Tierenaehr., Univ. Goettingen, Goettingen, Ger.).
42. McCabe, E., *Oxygen Therapies: A New Way of Approaching Disease*, Energy Publications, Morrisville, NY, 1988, p. 81.
43. Ibid.
44. Gainer, J., *Science News,* August 1971.
45. Carrel, A. and Ebeling, A.H., "Antagonistic Growth Principles of Serum and their Relation to Old Age, " *J. Exp. Med.*, vol. 38, 1923, pp. 419-25.

46. Simms, H.S. and Stillman, N.P., "Substances Affecting Adult Tissue in Vitro. II. A Growth Inhibitor in Adult Tissue," *J. Gen. Physiol.*, vol. 20, 1937, p. 603-19.
47. Hamaker, J.D., *The Survival of Civilization,* Hamaker-Weaver Publishers, Seymour, MO, 1982.
48. Levine, S.A. and Kidd, P.M., *Antioxidant Adaptation: Its Role in Free Radical Pathology*, Biocurrents Division, Allergy Research Group, San Leandro, CA, 1985, p. 38-41. Bradford, R.W., et al. *Oxidology: The Study of Reactive Oxygen Toxic Spores (ROTS) and their Metabolism in Health and Disease*, The Robert W. Bradford Foundation, Los Altos, CA, 1985.
49. Kleiner, I.S. and Orten, J.M., *Biochemistry*, 7th ed., C.V. Mosby, St. Louis, MO, 1966.
50. Cantarow, A. and Schepartz, B., *Biochemistry*, 4th ed., W.B. Saunders, Philadelphia, PA, 1967.
51. CSP, Symposium on DNA Replication and Recombination, Cold Spring Harbor, NY, June 1978. Reviewed in *Chem. & Engr. News*, June 19th and 26th, 1978.
52. Axmann, G., "Untersuchungen zur Organotropen Wirkung von Zellularen Extraken auf die Proteinsynthese in Vivo."
53. Gardner, M.L.G., "Intestinal Assimilation of Intact Peptides and Proteins from the Diet — A Neglected Field?" *Biol. Rev.*, vol. 59, 1984
54. Ibid.
55. Brambell, F.W.R., *The Transmission of Passive Immunity from Mother to Young*, Amsterdam Press, 1970.
56. Warshaw, A.L., Walker, W.A., and Isselbacher, K.J., "Transmission of Intact Protein in Humans," *J. Gastroenterology*, vol. 66, 1974, p. 987.
57. Bradford, R.W. and Allen, H.W., "The Absorption of Macromolecules, Including S.O.D. and Other Enzymes, Through the Mammalian Intestine," The Robert W. Bradford Research Inst., San Francisco, CA, 1984, p. 2.
58. Hemmings, W.A., ed., *Antigen Absorption by the Gut*, University Park Press, Baltimore, MD, 1978.
59. Richardson, J.S. et al., "Crystal Structure of Bovine Cu, Zn Superoxide Dismutase at $3A_0$. Resolution: Chain Tracing and Metal Ligands." *Proceedings of the National Academy of Sciences in the USA*, vol. 72, 1975, p. 1349.
60. Owen, R.L. and Nemaniac, P., *Scanning Electron Microscopy*, part 2, 1978, p. 367.
61. Owen, R.L., "Transport of Horseradish Peroxidase across the G-I Barrier," *Gastroent.*, vol. 72, 1977, p. 440.
62. Warshaw, A.L., et al., "Labeled Albumin Transport into the Lymphatic System," *Am. J. Surg.*, vol. 133, 1977, p. 55.
63. Hemmings, W.A. and Williams, E.W., "Food Antigens and Gut Transport," *Gut*, vol. 19, 1978, p. 715.
64. Hemmings, W.A., ed. *Antigen Absorption by the Gut.*
65. Kutsky, R.J., *Handbook of Vitamins, Minerals and Hormones*, 2nd ed., Van Nostrand Reinhold Co., NY, 1981, p. 179-295.
66. Vinson, J.A., Lecture Series, University of Scranton, Department of Chemistry, Scranton, PA., August 5th, 1981; May 10th, 1982; and June 23rd, 1982.
67. Baker, H. and Frank, O., College of Medicine and Dentistry of New Jersey, March 1987.

68. Gardner, M.L.G., "Intestinal Assimilation of Intact Peptides and Proteins from the Diet—A Neglected Field?"
69. Murachi, T., *Biochemical Aspects of Nutrition: The Proceedings of the First Congress of the Federation of Asian and Oceanian Biochemists.*
70. Murachi, T., Abstr. 10th International Congress of Biochemistry, Hamburg, Ger., 1976, p. 588.
71. Ambrus, M.D., et al., "Transport of Proteolytic Enzymes," *Clin. Pharmacol. Therap.*, vol. 8., 1967, p. 362.
72. Avakian, J., "Gastrointestinal Absorption of Intact Proteolytic Enzymes," *Clin. Pharmacol. Therap.*, vol. 5, 1964, p. 712.
73. Lehninger, A.L., *Biochemistry*, 2nd ed., Worth Publications, New York, NY, 1975, p. 59.
74. Owen, R.L., "Transport of Horseradish Peroxidase across the G-I Barrier."
75. Bloom, W. and Fawcett, D.W., *A Textbook of Histology*, W.B. Saunders Co., Philadelphia, PA, 1975, p. 394.
76. Zilliken, F. und Abdallah, K., *Molekularbiologische Grundlagen Des Kurz-und Langzeitgedachtnisses*, Stuttgart, Germany und New York, NY, 1973.
77. Griefe, H. and Molnar, S., (*Inst. Tierphysiol. Tierenaehr.*, Univ. Goettingen, Goettingen, Ger.). Z Tierphsiol., Tierernaehr, Futtermittelkd., Germany, vol. 40, no. 5, 1978, p. 248-56.
78. Frank, B.S., *Nucleic Acid Therapy of Aging and Degenerative Disease*, 3rd ed., Fiquima, Lisbon, 1975.
79. Lehninger, A.L., *Biochemistry*, p. 320.
80. Morris, I.G., "The Receptor Hypothesis of Protein Ingestion," in Hemmings, W.A., ed., *Antigen Absorption by the Gut*, Chapter 2, University Park Press, Baltimore, MD, 1978.
81. Lehninger, A.L., *Biochemistry*, 1975, p. 1002.
82. Williams, E.W. "Ferritin Uptake by the Gut of the Adult Rat: An Immunological and Electron-microscopic Study," in *Antigen Absorption by the Gut*, Hemmings, W.A., ed., Chapter 6, University Park Press, Baltimore, MD, 1978.
83. Murachi, T., "Intestinal Absorption of Enzyme Proteins," in *Biochemical Aspects of Nutrition, the Proceedings of the First Congress of the Federation of Asian and Oceanian Biochemists*, Yagi, K., ed., Japan Scientific Societies Press, Tokyo. American distributor: University Park Press, Baltimore, MD, 1979.
84. Gardener, M.L.G., "Intestinal Assimilation of Intact Peptides and Proteins from the Diet—A Neglected Field?"
85. Kalb, H.W., "Ober die Spezifisch Stoffwechselsteigernde Wirkung von Organextrakten in Vitro." Inaugural-Dissertation aus dem Path. Inst. d. Univ. Munchen, 1959.
86. Wacker, A., "Beeinflussung der Proteinsynthese durch in Vivo-Verabreichung von Organspezifischen RNS-Praparaten." *Personliche Mitteilung* vom 7.7.71 (unveroffentlichte Versuche). Published elsewhere from experiments he conducted at the Institute for Therapeutical Biochemistry at the Frankfurt/Main University.
87. Cameron, D.E., "The Use of Nucleic Acid in Aged Patients with Memory Impairment," *Amer. J. Psychiat.*, vol. 114, 1958, p. 943.

88. Dyckerhoff, H., *Monographic Information about Ribonucleic Acids: Regeneresen*, Muller\Goppingen, Chemisch-Pharmazeutische Fabrik, Cologne, Federal Republik of Germany, p. 11.

89. Newman, E.A., et al., "Effect of Nucleic Acid Supplements in the Diet on Rate of Regeneration of Liver in Rats."

90. Rothschild, P.R. *Enzyme-Therapy in Immune Complex and Free Radical Contingent Diseases*, University Labs Press, Honolulu, HI, 1988.

91. Greife, H. and Molnar, S., "Studies of Nucleic Acid Metabolism in Rats by use of Carbon-14-Labeled Purine and Pyrimidine Bases and Nucleic Acids. Anabolic Pathways of Nucleic Acid Derivatives," *Z. Tierphysiol.*, Tierernaehr. Futtermittelkd, vol. 40, no. 5, 1978, p. 248-56, Univ. Goettingen, Goettingen, Ger.

92. Kugler, H.J., *Slowing down the Aging Process*, Pyramid Publications, New York, NY, 1976, p. 193-94.

93. Krementz, E.T. and Hornung, M.O., "Specific Tissue and Tumor Responses of Chimpanzees Following Immunization against Human Melanoma," *Surgery*, vol. 75, 1974, p. 477-86.

94. Odens, "Prolongation of the Life Span in Rats," *J. Amer. Geriatr. Soc.*, vol. 21, no. 10, 1973, p. 450-51.

95. Louisot, P. and Colobert, L., "Inhibition de la Multiplication Virale a l'Aide d'Acides Ribonucleiques chimiquement Modifies," *Biochim. Biophys. Acta*, vol. 155, 1968, p. 38-50.

96. Ebel, J.P., et al., "Inhibition of the Multiplication of the Myxovirus and Arbovirus by Chemically Modified Ribonucleic Acids from the Host Cells," *Biochem. Biophys. Res. Comm.*, vol. 30, 1968, p. 148-52.

97. Tikhonenko, T.I., et al., "Inhibitory Effect of Normal Cell RNA on Virus Multiplication," *Fed. Proc.*, vol. 23, 1964, p. 998-1002.

98. Sugahara, T., et al., "Effect of an Alkaline-Hydrolyzed Product of Yeast RNA on the Survival of Repeatedly Irradiated Mice," *Radiation Research*, vol. 29, 1966, pp. 516-22.

99. Maisin, J., et al., "Yeast Ribonucleic Acid and its Nucleotides as Recovery Factors in Rats Receiving an Acute Whole-Body Dose of X-rays," *Nature*, vol. 186, 1960, p. 475-87.

100. Ebel, J.P., et al., "Study of the Therapeutic Effect on Irradiated Mice of Substances Contained in RNA Preparations," *Int. J. Radiat. Biol.*, vol. 16, 1969, p. 201-09.

101. Wagner, R. and Silverman, E.C., "Chemical Protection against X-radiation in the Guinea-Pig. I. Radioprotective Action of RNA and ATP," *Int. J. Rad. Biol.*, vol. 12, 1967, p. 101-12.

102. Batkin, S., "The Effect of RNA on the Healing of Carp Spinal Cord," *Proc. Nat. Acad. Sci. USA*, vol. 56, 1966, p. 1689-91.

103. Belous, A.M., et al., "Effect of Exogenous RNA and Ultrasound on Fracture Healing in Rats," *Byull. Eksp. Biol. Med.*, vol. 67, 1969, pp. 85-88. USSR (See also *Pharmacology*, vol. 110, 1971, p. 890.)

104. Williamson, M.B. and Guschlbauer, W., "Metabolism of Nucleic Acids during Regeneration of Wound Tissues II., The Rate of Formation of RNA," *Arch. Biochem. Biophys.*, vol. 100, 1963, p. 245-50.

105. Groth, C.G., et al., "Effect of Ribonucleic Perfusion on Canine Kidney and Liver Homograph Survival," *Surgery*, vol. 64, 1968, p. 13-38.
106. Fuller, R.W., et al., "Serum Uric Acid in Mongolism," *Science*, vol. 137, 1962, p. 868-69.
107. Enesco, H.E., "RNA and Memory – A Re-Evaluation of Present Data," *Canad. Psychiat. Ass. J.*, vol. 12, 1967, p. 29-34.
108. Bear, F.E., "Earth: The Stuff of Life," University of Oklahoma Press, Stillwater, OK, 1962.
109. Beach, Rex, *Modern Miracle Men*, 74th Congress, 2nd Session, U.S. Senate Document No. 264, June 1, 1936, United States Government Printing Office, Washington, DC, 1941, p. 1.
110. Cell Tech, *The Miracle of Super Blue Green Algae*. Klamath Falls, OR, p. 36.
111. National Institute of Health, *How to Cope with Arthritis*, US Department of Health and Human Services, Public Health Service, NIH Publications No. 82-1092, October 1991.
112. Fry, T.C., "The Myth of Health in American," *Dr. Shelton's Hygenic Review*, vol. 37, no. 7, 1976, p. 150-52.
113. Ibid.
114. Kollman, D., *Hope Is A Molecule,* Cell Tech, Klamath Falls, OR, 1989, p. 2.
115. "Nature's Most Perfect Food," *Creative Living*, August 1990, p. 19.
116. Clement, G., et al., "Amino Acid Composition and Nutritive Value of the Alga Spirulina Maxima," *J. Sci., Food, and Agr.*, vol. 18, no. 11, p. 497-501.
117. Promotional literature from Metagenics/Ethical Nutrients.
118. Evans, W.J., et al., "Changes in Whole Body Leucine Dynamics During Submaximal Exercise in Human Subjects," *Med. Sci. Sports*, vol. 12, 1981, p. 89.
119. Yamada, Y., et al., "School Children's Growth and the Value of Chlorella," *Nihon iji Shimpo*, 1966, p. 2196.
120. "The Administration of Chlorella Extract, and the Physique and Physical Fitness of Elementary School Pupils," Department of Hygiene, Nagassaki University Medical School.
121. Meguid, M.M., et al., "Effect of Elemental Diet on Albumin and Urea Slushiest: Comparison with Partially Hydrolyzed Protein Diet," *Journal of Surgical Research*, vol. 37, 1984, p. 16-24.
122. Drapeau, C., *Physiology of the Blue Green Algae*, Cell Tech, Klamath Falls, OR, 1994, p. 3.
123. Rothschild, P.E. and Fehey, W.J. "Free Radicals, Stress and Antioxidant Enzymes: A Guide to Cellular Health," third edition, University Labs Press, Honolulu, HI, 1991.
124. Dimitriadis, G.J., "Introduction of Ribonucleic Acids into Cells by Means of Liposomes."
125. Personal correspondence, Cell Tech letter to Dr. John Apsley, February 1992.
126. CSP, Symposium on DNA Replication and Recombination, Cold Spring Harbor, NY, June 1978. Reviewed in *Chem.* and *Engr. News*, June 19th and 26th, 1978.
127. Apsley, J.W., "Blue Green Algae and the Factors of Regeneration," Genesis Communications, LLC, Northport, AL, 1995.
128. Pert, C.B., "The Wisdom of the Receptors: Neuropeptides, the Emotions, and Bodymind," *Advances*, vol. 3, no. 3, 1986.

129. Hagino, et al., "Effect of Chlorella on Fecal and Urinary Cadmium Excretion in `itai'itai." *Japanese J. Hyg.*, vol. 30, no. 1, 1975, p. 77.
130. Pore, R.S., "Detoxification of Chlordecon Poisoned Rats with Chlorella and Chlorella Derived Sporopollenin," *Drug Chem. Toxicol.*, vol. 7, no. 1, 1984, p. 57-71.
131. Yamane, Y., "The Effect of Spirulina on Nephrotoxicity in Rats," from a paper presented at the Pharmaceutical Society of Japan's Annual Symposium, April 15, 1988.
132. Horikoshi, T.A., et al., "Uptake of Uranium by Various Cell Fractions of Chlorella Regularis," *Radioisotopes,* vol. 28, no. 8, 1978, p. 485-87.
133. Qishen, P., et al., "Radioprotective Effect of Extract from Spirulina Platensis in Mouse Bone Marrow Cells Studied by Using the Micronucleus Test," *Toxicology Letters*, vol. 48, 1989, p. 165-69.
134. Kay, R.A., "Microalgae as Food and Supplement," *Clinical Reviews in Food Sci. and Nutr.*, vol. 30, 1991, p. 566.
135. Lopez-Romero, D., "Gamma Linolenic Acid as a Base of Treatment for Chronic Infirmities. Clinical Experience in Spain with Evening Primose Oil and Spirulina Microalgae," from a paper presented at Medicina Holistica, Madrid, Spain, Oct. 12, 1987.
136. Beach, Rex, *Modern Miracle Men*, p. 1.
137. *The Miracle Of Super Blue Green Algae*, Cell Tech, Klamath Falls, OR, 1994, p. 35.
138. Gustafson, K.R., et al., "AIDS-Antiviral Sulfolipids From Cyanobacteria (Blue-Green Algae)," *J. Natl. Cancer Inst.*, vol. 81, 1989, p. 1254-58.
139. Drapeau, C., *Physiology of the Blue Green Algae*, p. 4.
140. Konishi, R.K., et al., "Antitumor Effect Induced by a Hot Water Extract of Chlorella Vulgaris: Resistance to Meth-a-Tumor Growth Mediated by CE-Induced Polymorphonuclear Leukocytes," *Cancer Immunol. Immunother.*, vol. 19, 1985, p. 73-78.
141. Matsueda, S., et al., "Studies on Antitumor Active Glycoprotein from Chlorella Vulgaris," *Yajugaku Zasshi*, vol. 102, 1982, p. 447-51.
142. "Nature's Most Perfect Food," *Creative Living*, August 1990, p. 19.
143. Challam, J., *Spirulina*, p. 3 and 6.
144. "Green Giants – Ancient Algae and Modern Cereal Grasses. Spirulina: Heavenly Nutrient," *Delicious!*, July/August 1990, p. 34-35.

INDEX

New from Genesis Communications

The Genesis Effect:

Spearheading Regeneration with Wild Blue Green Algae

by Dr. John W. Apsley

An interesting and scientifically-documented explanation of how algae initiates regeneration within the human body!

Wake up now, pay attention — this book will dramatically change your life. Read diligently to profoundly help yourself, your family and your friends find health secrets that should never have been secrets at all. In reading the story of *The Genesis Effect* and blue green algae you are embarking upon the most important journey of your life: the discovery of how to take charge over the true nutritional content of the food that enters your body so that you can begin to live a healthier, fuller life.

- You will learn about regeneration (your body's ability to begin healing itself when properly nourished)!
- You will learn why so few people today are truly healthy. The U.S. Public Health Department has stated that only 1 1/2% of the population is truly healthy.
- You will learn how to begin the process of regeneration within your own body.

The human body has one ability not possessed by any machine – the ability to repair itself.

Charles E. Kriley, Jr., M.D

Book Order Form

Genesis Communications, LLC
1514 Skyland Blvd. E., Suite #283
Tuscaloosa, AL 35405
(205) 556-6479 or (205) 333-9788
(800) 772-4596
Internet: 74561.3705@compuserve.com

Ordered by:

____
Name Date

____
Street Address

____
City State Zip

____
Phone Fax

___ Check or Money Order ___ VISA ___ MasterCard

__________________________ __________________
Card Number Expiration Date

Signature____

Please send me _____ books. I am enclosing $__________ ($7.95 per book) plus shipping (see below). Send check or money order. No cash or C.O.D.'s please.

Send orders to: Genesis Communications, LLC, 1514 Skyland Blvd. E., Suite #283, Tuscaloosa, AL 35405 or call us at the numbers listed above.

Shipping and Handling Charges

	UPS Ground or US Mail	**Priority Mail**	**UPS Express**
1st Book	x $3.50	x $4.50	x $9.00
Each Additional Book	x $1.00	x $0.50	x $1.00

Volume discounts are available. Please call for information.

For information on how to order Wild Blue Green Algae, please contact the person who sold or gave you this book!!